Hands-On Robotics Programming with C++

Leverage Raspberry Pi 3 and C++ libraries to build intelligent robotics applications

Dinesh Tavasalkar

BIRMINGHAM - MUMBAI

Hands-On Robotics Programming with C++

Copyright © 2019 Packt Publishing

Commissioning Editor: Vijin Boricha
Acquisition Editor: Shrilekha Inani
Content Development Editor: Abhijit Sreedharan
Technical Editor: Mohd Riyan Khan
Copy Editor: Safis Editing
Language Support Editor: Mary McGowan
Project Coordinator: Jagdish Prabhu
Proofreader: Safis Editing
Indexer: Priyanka Dhadke
Graphics: Tom Scaria
Production Coordinator: Jisha Chirayil

First published: March 2019

Production reference: 1290319

Published by Packt Publishing Ltd.
Livery Place
35 Livery Street
Birmingham
B3 2PB, UK.

ISBN 978-1-78913-900-6

www.packtpub.com

This book is dedicated to my mentor, Asha Sundararajan. It was under her guidance that I began my career in robotics.

`mapt.io`

Mapt is an online digital library that gives you full access to over 5,000 books and videos, as well as industry leading tools to help you plan your personal development and advance your career. For more information, please visit our website.

Why subscribe?

- Spend less time learning and more time coding with practical eBooks and Videos from over 4,000 industry professionals

- Improve your learning with Skill Plans built especially for you

- Get a free eBook or video every month

- Mapt is fully searchable

- Copy and paste, print, and bookmark content

Packt.com

Did you know that Packt offers eBook versions of every book published, with PDF and ePub files available? You can upgrade to the eBook version at `www.packt.com` and as a print book customer, you are entitled to a discount on the eBook copy. Get in touch with us at `customercare@packtpub.com` for more details.

At `www.packt.com`, you can also read a collection of free technical articles, sign up for a range of free newsletters, and receive exclusive discounts and offers on Packt books and eBooks.

Contributors

About the author

Dinesh Tavasalkar is a trainer and online instructor from India. He has trained more than 8,000+ students on topics related to robotics, Internet of Things, Arduino, Raspberry Pi, Android app development, augmented reality, and virtual reality. Dinesh's online courses have been undertaken by 25,000+ people on Udemy from more than 150+ countries. Some of his popular courses on Udemy include *Robotics for beginners, Smartphone Control Robot using Arduino and Android, Build Augmented Reality apps using Unity and Vuforia,* and *Build Virtual Reality games for Google Cardboard using Unity.* He also runs a YouTube channel called *EngineersDream,* where he teaches Android application development.

About the reviewers

Lentin Joseph is an author and robotics entrepreneur from India. He runs a robotics software company called Qbotics Labs in India. He has 8 years' experience in the robotics domain, primarily in ROS, OpenCV, and PCL. He has authored seven books on ROS, namely *Learning Robotics Using Python*, (First and Second Editions), *Mastering ROS for Robotics Programming*, (First and Second Editions), *ROS Robotics Projects*, *Robot Operating System for Absolute Beginners*, and *ROS Programming*. He did his master's in robotics in India and also did some research at Robotics Institute, CMU, USA.

Shahid Memon holds an MSc in autonomous robotics engineering and possesses a BSc in Computer Science. He has collaborated with colleagues on product feasibility studies and new product ideas to meet clients' needs and support companies' objectives. He has coordinated several product development projects and assisted in the design and testing phases. He is a strategic thinker with the ability to drive company goals and analyze research to impact product and business needs. He is an avid researcher of the latest trends within the technology industry and how these trends affects business. He is a proven leader with outstanding communication, interpersonal, project management, and supervisory skills.

Packt is searching for authors like you

If you're interested in becoming an author for Packt, please visit `authors.packtpub.com` and apply today. We have worked with thousands of developers and tech professionals, just like you, to help them share their insight with the global tech community. You can make a general application, apply for a specific hot topic that we are recruiting an author for, or submit your own idea.

Table of Contents

Preface

C++ is one of the most popular legacy programming languages for robotics, and a combination of C++ and robotics hardware is used in many leading industries. This book will bridge the gap between Raspberry Pi and C/C++ programming and enable you to develop applications for Raspberry Pi. To follow along with the projects covered in the book, you can implement C programs in Raspberry Pi with the wiringPi library.

With this book, you'll develop a fully functional car robot and write programs to move it in different directions. You'll then create an obstacle-avoiding robot using an ultrasonic sensor. Furthermore, you'll find out how to control the robot wirelessly using your PC/Mac. This book will also help you work with object detection and tracking using OpenCV, and guide you through exploring face detection techniques. Finally, you will create an Android app and control the robot wirelessly with an Android smartphone.

By the end of this book, you will have gained experience in developing a robot using Raspberry Pi and C/C++ programming.

Who this book is for

This book is designed for developers, programmers, and robotics enthusiasts interested in leveraging C++ to build exciting robotics applications. Some prior knowledge of C++ is necessary.

What this book covers

Chapter 1, *Introduction to the Raspberry Pi*, covers different modes of Raspberry Pi and GPIO pin configuration. Then, we will set up Raspberry Pi B+ and Raspberry Pi Zero and install the Raspbian OS on it. We will also learn how to connect a Raspberry Pi to a laptop wirelessly via a Wi-Fi network.

Chapter 2, *Implementing Blink with wiringPi*, covers the installation of the wiringPi library. In this chapter, we will understand the wiringPi pin connections for the Raspberry Pi. Then, we will write two C++ programs and will upload them onto our Raspberry Pi.

Chapter 3, *Programming the Robot*, covers the criteria for selecting a robot chassis. After that, we will construct our car, connect the motor driver to the Raspberry Pi, and understand the workings of an H-bridge circuit. Finally, we will write programs to move the robot forward, backward, left, and right.

Chapter 4, *Building an Obstacle-Avoiding Robot*, looks at how an ultrasonic sensor works, and we will write a program to measure distance values. Next, we will program the 16 x 2 LCD to read the ultrasonic distance value. We will also look at the I2C LCD, which takes the 16 LCD pin as an input and provides four pins as an output, thus simplifying the wiring connections. Finally, we will fit the ultrasonic sensor on our robot to create our obstacle-avoiding robot. This robot will move freely when there are no obstacles near it, and if it approaches an obstacle, it will avoid it by taking a turn.

Chapter 5, *Controlling a Robot Using a Laptop*, looks at two different techniques for controlling the robot using a laptop. In the first technique, we will use the ncurses library to take input from the keyboard to move the robot accordingly. In the second technique, we will use the QT Creator IDE to create GUI buttons, and then use these buttons to move the robot in different directions.

Chapter 6, *Accessing Rpi Cam with OpenCV*, focuses on the installation of OpenCV on the Raspberry Pi. You will also be introduced to the Raspberry Pi camera module and, after setting up the Pi camera, you will take pictures and record a short video clip using the Pi camera.

Chapter 7, *Building an Object-Following Robot with OpenCV*, covers some of the important functions inside OpenCV libraries. After that, we will put these functions to the test and attempt to recognize an object from an image. Then, we will learn how to read a video feed from the Pi camera, how to threshold a colored ball, and how to place a red dot on top of it. Finally, we will use the Pi camera and the ultrasonic sensor to detect the ball and follow it.

Chapter 8, *Face Detection and Tracking Using Haar Classifier*, uses the Haar face classifier to detect a face from a video feed and draw a rectangle around it. Next, we will detect eyes and a smile on the given face and create a circle surrounding the eyes and mouth. After using this knowledge of face and eye detection, we will first turn the LED on/off when the eyes and smile are detected. Next, by creating a white dot in the center of the face, we will make the robot follow the face.

`Chapter 9`, *Building a Voice-Controlled Robot*, starts with creating our first Android application, called Talking Pi, in which text written inside the textbox will be displayed in a label and also read out by the smartphone. We will then develop a voice-controlled Android app for the bot, which will recognize our voice and send text to the RPi via Bluetooth. After this, using the terminal window, we will pair the Android smartphone's Bluetooth with the RPi's Bluetooth. Finally, we will look at socket programming and write the VoiceBot program to establish a connection with the Android smartphone's Bluetooth in order to control the robot.

To get the most out of this book

To get through the code in this book, Raspberry Pi 3B+ or Raspberry Pi Zero board is required. The additional hardware and software is mentioned in the *Technical requirements* section of each chapter.

Download the example code files

You can download the example code files for this book from your account at `www.packt.com`. If you purchased this book elsewhere, you can visit `www.packt.com/support` and register to have the files emailed directly to you.

You can download the code files by following these steps:

1. Log in or register at `www.packt.com`.
2. Select the **SUPPORT** tab.
3. Click on **Code Downloads & Errata**.
4. Enter the name of the book in the **Search** box and follow the onscreen instructions.

Once the file is downloaded, please make sure that you unzip or extract the folder using the latest version of:

- WinRAR/7-Zip for Windows
- Zipeg/iZip/UnRarX for Mac
- 7-Zip/PeaZip for Linux

The code bundle for the book is also hosted on GitHub at `https://github.com/PacktPublishing/Hands-On-Robotics-Programming-with-Cpp`. In case there's an update to the code, it will be updated on the existing GitHub repository.

We also have other code bundles from our rich catalog of books and videos available at https://github.com/PacktPublishing/. Check them out!

Download the color images

We also provide a PDF file that has color images of the screenshots/diagrams used in this book. You can download it here: http://www.packtpub.com/sites/default/files/downloads/9781789139006_ColorImages.pdf.

Conventions used

There are a number of text conventions used throughout this book.

CodeInText: Indicates code words in text, database table names, folder names, filenames, file extensions, pathnames, dummy URLs, user input, and Twitter handles. Here is an example: "The code for taking axial and radial turns is added to the RobotMovement.cpp program."

A block of code is set as follows:

```
digitalWrite(0,HIGH);          //PIN 0 & 2 will STOP the Left Motor
digitalWrite(2,HIGH);
digitalWrite(3,HIGH);          //PIN 3 & 4 will STOP the Right Motor
digitalWrite(4,HIGH);
delay(3000);
```

When we wish to draw your attention to a particular part of a code block, the relevant lines or items are set in bold:

```
digitalWrite(0,HIGH);          //PIN 0 & 2 will STOP the Left Motor
digitalWrite(2,HIGH);
digitalWrite(3,HIGH);          //PIN 3 & 4 will STOP the Right Motor
digitalWrite(4,HIGH);
delay(3000);
```

Any command-line input or output is written as follows:

```
sudo nano /boot/config.txt
```

Bold: Indicates a new term, an important word, or words that you see on screen. For example, words in menus or dialog boxes appear in the text like this. Here is an example: "Select the **Remember password** option and press **OK**."

Warnings or important notes appear like this.

 Tips and tricks appear like this.

Get in touch

Feedback from our readers is always welcome.

General feedback: If you have questions about any aspect of this book, mention the book title in the subject of your message and email us at customercare@packtpub.com.

Errata: Although we have taken every care to ensure the accuracy of our content, mistakes do happen. If you have found a mistake in this book, we would be grateful if you would report this to us. Please visit www.packt.com/submit-errata, selecting your book, clicking on the Errata Submission Form link, and entering the details.

Piracy: If you come across any illegal copies of our works in any form on the internet, we would be grateful if you would provide us with the location address or website name. Please contact us at copyright@packt.com with a link to the material.

If you are interested in becoming an author: If there is a topic that you have expertise in, and you are interested in either writing or contributing to a book, please visit authors.packtpub.com.

Reviews

Please leave a review. Once you have read and used this book, why not leave a review on the site that you purchased it from? Potential readers can then see and use your unbiased opinion to make purchase decisions, we at Packt can understand what you think about our products, and our authors can see your feedback on their book. Thank you!

For more information about Packt, please visit packt.com.

Section 1: Getting Started with wiringPi on a Raspberry Pi

In this section, you will first be introduced to the basics of Raspberry Pi and learn how to install the Raspbian OS on your Raspberry Pi. Next, you will work with the wiringPi library and execute your first C program on Raspberry Pi.

The following chapters are included in this section:

- Chapter 1, *Introduction to the Raspberry Pi*
- Chapter 2, *Implementing Blink with wiringPi*

Introduction to the Raspberry Pi 1

Initially developed with the idea of teaching and promoting basic computer programming in schools across the UK, the **Raspberry Pi** (**RPi**) became an instant hit. At a price of just $25 when it was initially released, it became so popular that it was, and still is, used by developers, hobbyists, and engineers all over the world.

In this chapter, you will explore the basic idea of a Raspberry Pi. You will then learn to install an operating system on the device. Finally, you will configure Wi-Fi on your Raspberry Pi and learn to connect it to a laptop over Wi-Fi and set up a remote desktop.

You will achieve each of these objectives through the following topics:

- Understanding the Raspberry Pi
- Installing Raspbian OS on a Raspberry Pi 3B+
- Connecting a Raspberry Pi 3B+ to a laptop via Wi-Fi
- Installing Raspbian OS on a Raspberry Pi Zero W
- Connecting a Raspberry Pi Zero W to a laptop via Wi-Fi

Technical requirements

For this chapter, the following software and hardware will be required.

Software required

Please download the following software if you want to follow along with the instructions in this chapter:

- **Raspbian Stretch**: Raspbian Stretch is the **operating system** (**OS**) that we will write to a microSD card. Stretch is the OS that will run our Raspberry Pi. It can be downloaded from `https://www.raspberrypi.org/downloads/raspbian/` This OS is developed specifically for the Raspberry Pi.
- **Balena Etcher**: This software will format the microSD card and write the Raspbian Stretch image to the microSD card. It can be downloaded from `https://www.balena.io/etcher/`.
- **PuTTY**: We will use PuTTY to connect our Raspberry Pi to a Wi-Fi network and find the IP address that the Wi-Fi network assigns to it. It can be downloaded from `https://www.chiark.greenend.org.uk/~sgtatham/putty/latest.html`.
- **VNC Viewer**: With VNC Viewer, we will be able to view the Raspberry Pi display on our laptop. It can be downloaded from `https://www.realvnc.com/en/connect/download/viewer/`.
- **Bonjour**: This is generally used to connect printers to computers over Wi-Fi. It can be downloaded from `https://support.apple.com/kb/DL999?viewlocale=en_MYlocale=en_MY`.
- **Notepad++**: We will need Notepad++ to edit the code in the Raspbian Stretch image. It can be downloaded from `https://notepad-plus-plus.org/download/v7.5.9.html`.
- **Brackets:** Brackets allows those using macOS to edit the code in the Rapbian Stretch image. To download Brackets, go to `http://www.brackets.io/`.

The installation of all of this software is pretty straightforward. Keep the default settings checked, click on the **Next** button a few times, and then hit the **Finish** button once the installation is complete.

Hardware requirements

We need the following hardware to follow along with the instructions in this chapter.

For Raspberry Pi 3B+ and Raspberry Pi Zero W

If you use the Raspberry Pi 3B+ or the Raspberry Pi Zero W, you will need the following hardware:

- Keyboard
- Mouse
- SD card—this should have a minimum of 8 GB of storage, but 32 GB is recommended
- MicroSD card reader
- Display unit—a computer monitor or TV that contains an HDMI port
- HDMI cable
- 5V mobile charger or power bank. This will power the Raspberry Pi

Additional hardware for Raspberry Pi 3B+

The Raspberry Pi 3B+ needs the following additional hardware:

- An Ethernet cable

Additional hardware requirements for Raspberry Pi Zero W

Since the Raspberry Pi Zero has a micro USB port and a Micro HDMI port, it needs the following additional hardware:

- A USB hub
- A micro USB B-to-USB connector (also known as an OTG connector)
- An HDMI-to-mini HDMI connector

Understanding the Raspberry Pi

Raspberry Pi is a credit card sized, Linux-based minicomputer invented by the Raspberry Pi Foundation in 2012. The first Raspberry Pi model was called the Raspberry Pi 1B, which was then followed by the Model A. Raspberry Pi boards were initially intended to promote computer science programs in schools. However, their inexpensive hardware and free, open source software, quickly made the Raspberry Pi popular among hackers and robotics developers.

The Raspberry Pi can be used as a fully functional computer. It can be used to perform tasks such as browsing the internet, playing games, and watching HD videos, as well as creating Excel and Word documents. But what really differentiates it from a normal computer is its programmable GPIO pins. The Raspberry Pi consists of **40 digital I/O GPIO pins** that can be programmed.

In simple terms, the Raspberry Pi can be thought of as a combination of a **minicomputer**, as it can be used as a fully fledged computer, and an **electronics hardware board**, as it can be used to create electronics and robotics projects.

There are different Raspberry Pi models. In this book, we are going to be using the following two models:

- The Raspberry Pi 3B+
- The Raspberry Pi Zero W

The Raspberry Pi 3B+

The Raspberry Pi 3B+ was released in February 2018. Its specifications are shown in the following annotated photo:

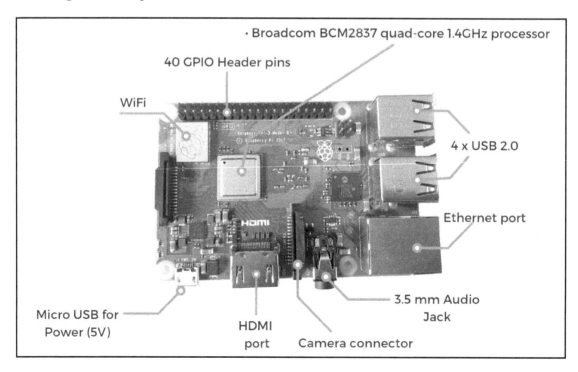

The Raspberry Pi 3B+ specifications are as follows:

- A Broadcom BCM2837 quad-core 1.4 GHz processor
- 1 GB RAM
- A Broadcom VideoCore GPU
- Bluetooth 4.2
- Dual-band 2.4 GHz and 5 GHz Wi-Fi
- An Ethernet port
- Storage with a microSD card via a microSD slot
- 40 programmable GPIO pins
- Four USB 2.0 ports
- An HDMI port
- A 3.5 mm audio jack
- The **Camera Serial Interface** (**CSI**), used for connecting the Raspberry Pi Camera directly to the Raspberry Pi

The Raspberry Pi Zero W

If we were looking for a smaller-sized version of the Raspberry Pi, we could instead opt for the Raspberry Pi Zero W. The **W** stands for **wireless**, as the Raspberry Pi Zero W has built-in Wi-Fi. The following is an annotated photo of the Raspberry Pi Zero W:

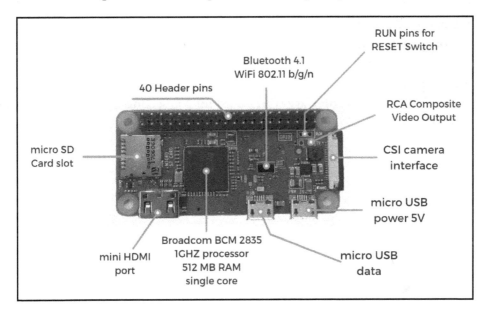

The Raspberry Pi Zero W model costs around $10. There is also Raspberry Pi Zero, without the **W**, which costs around $5, but this does not have built-in Wi-Fi, which makes it very difficult to connect it to the internet. The Raspberry Pi Zero W, which was released in 2017, is basically an updated version of the Raspberry Pi Zero, released in 2015.

 Later on in this book, when we design our robot, we will learn how to upload a program wirelessly to our Raspberry Pi from a laptop via a Wi-Fi network. If you opt to purchase the smaller version of Raspberry Pi, I recommend that you go with Raspberry Pi Zero W instead of Raspberry Pi Zero, for ease of use.

The Raspberry Pi Zero W has a couple of disadvantages, due to its small size. Firstly, it's a bit slower than the Raspberry Pi 3B+. Secondly, if we want to use it as a minicomputer, we would need to purchase different extensions to connect peripherals such as a keyboard, a mouse, or a monitor. If we are going to use the Raspberry Pi Zero W for building electronics and robotics projects, however, we don't need to worry about this drawback. Later on in this book, we will learn how to connect the Raspberry Pi Zero W to a laptop via Wi-Fi and how to use the laptop to control it as well.

The specifications of the Raspberry Pi Zero W are as follows:

- A Broadcom ARM11 1 GHz processor
- 512 MB RAM
- A Broadcom VideoCore GPU
- Bluetooth 4.0
- Dual-band 2.4 GHz and 5 GHz Wi-Fi
- Storage with a microSD card via a microSD slot
- 40 programmable GPIO pins
- A mini HDMI port
- The **Camera Serial Interface** (**CSI**), used for connecting the Raspberry Pi Camera directly to the Raspberry Pi

Setting up a Raspberry Pi 3B+ as a desktop computer

To set up the Raspberry Pi 3B + and install the Raspbian OS on it, we will need various hardware and software components. The hardware components include the following:

- A laptop to install Raspbian OS on a microSD card.
- A keyboard.
- A mouse.
- An SD card—a minimum of an 8 GB memory card is more than sufficient, but with an 8 GB card, the default OS will occupy 50 percent of memory card space. Later on in this chapter, we will also install OpenCV on your Raspberry Pi, and since OpenCV will also occupy a lot of space on your memory card, you will need to uninstall some default software. So, I recommend you use a 16 GB or 32 GB memory card—with a 32 GB memory card, the default OS only occupies 15 percent of the card's space.
- An SD card reader.
- A display unit—this can be a computer monitor or TV, as long as it features an HDMI port.
- An HDMI cable.
- A mobile charger or a power bank to power the Raspberry Pi.

The software components required include the following:

- Etcher
- Raspbian Stretch with Desktop OS

Now that we know what we need to install the OS, let's start installing it.

Installing Raspbian OS on an SD card

To install Raspbian OS on a microSD card, we will first install **Etcher** on our computer. After that, we will insert the microSD card into a microSD card reader and connect it to our computer.

Downloading and installing Etcher

Etcher will first format the microSD card, and then write the Raspbian Stretch image to it. Let's start by installing Etcher:

1. On your browser, go to `http://www.etcher.io/`.
2. Select your OS from the drop-down menu. Etcher will start downloading, as shown in the following screenshot:

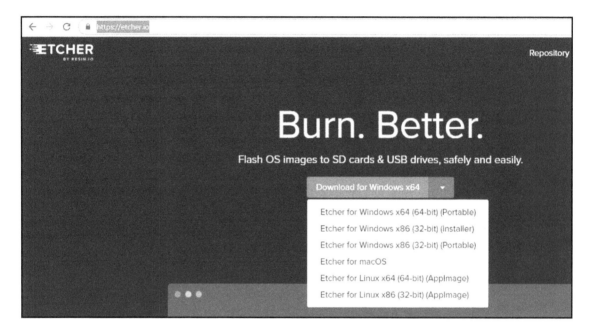

3. Once it has downloaded, open the setup file and install Etcher.

Now that Etcher is set up, let's move on to Raspbian.

Downloading the Raspbian Stretch image

We now have to download an OS to run on our Raspberry Pi. While there are many third-party Raspberry Pi OSes available, we will install Raspbian OS. This OS is based on Debian and is developed specifically for the Raspberry Pi. The newest version is called **Raspbian Stretch**.

To download the Raspbian Stretch image, visit https://www.raspberrypi.org/downloads/ raspbian/, look for the **RASPBIAN STRETCH WITH DESKTOP** ZIP file, and click on the **Download ZIP** button, as shown in the following screenshot:

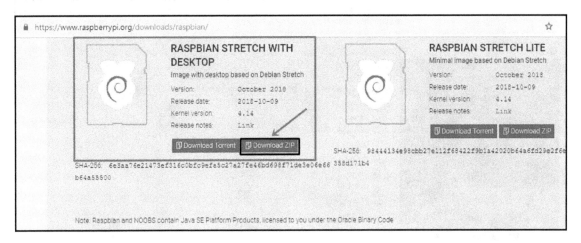

Now that we have a copy of Raspbian Stretch on our laptops, let's move on to writing it to our microSD cards.

Writing the Raspbian Stretch image to a microSD card

After downloading Etcher and the Raspbian Stretch image, let's write Raspbian Stretch to our microSD card:

1. Insert the microSD card into the microSD card reader, then connect the card reader to your laptop via USB:

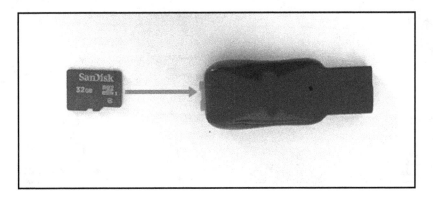

2. Next, open Etcher and click on the **Select Image** button. After this, select the Raspbian Stretch ZIP file and click on **Open**:

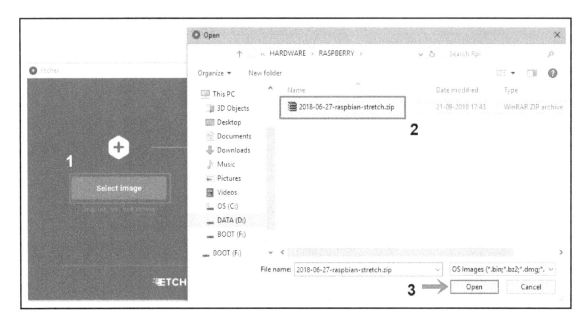

3. After that, make sure that the microSD card reader drive is selected, as in the following screenshot. If any other drive is selected by mistake, click on the **Change** button and select the microSD card drive. Click on the **Flash!** button to write the Raspbian OS to the microSD card:

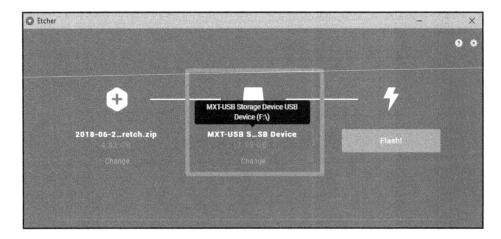

The process of writing or flashing an image to an SD card is also called **booting**.

It will take around 15-20 minutes for Etcher to flash your SD card with the Raspbian OS:

Once the OS is written to the SD card, Etcher will automatically eject the microSD card reader.

Now that we've written Raspbian Stretch to our microSD card, let's begin setting up the Raspberry Pi 3B+.

Setting up the Raspberry Pi 3B+

After booting the Raspbian OS from the microSD card, we will set up the Raspberry Pi by connecting different peripherals to it, as follows:

1. Insert the microSD card into the SD card slot, which is located on the back of the Raspberry Pi 3B+:

2. Connect the keyboard and mouse to the USB ports of the Raspberry Pi 3B+. It is also possible to use a wireless keyboard and mouse:

3. The Raspberry Pi 3B+ contains an HDMI port with which we can connect the RPi to a display unit, such as a computer monitor or TV. Connect one end of the HDMI cable to the Raspberry Pi's HDMI port and the other end to a display unit:

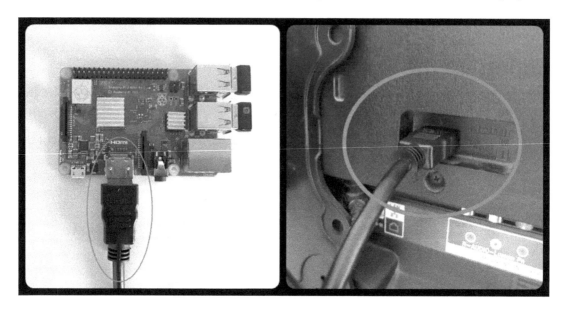

4. Finally, to turn the Raspberry Pi on, we need to provide it with power. A typical Raspberry Pi requires 5V of power and ideally 2.5A of current. There are two methods that we can use to supply power to the Raspberry Pi:

- **Smartphone charger**: Most smartphone chargers provide a 5V voltage output and 2.5A of current output. If you take a closer look at your smartphone charger, you will find the maximum voltage and the current output value printed on it, as shown in the following photo. On my charger, the current output of 3A indicates the maximum current output. The charger, however, will only provide the current output as required for the RPi and not the maximum current of 3A. Note that the Raspberry Pi contains a micro **USB B** port, so, to connect to the power port of Raspberry Pi, we need to connect a micro **USB B** wire to our charger:

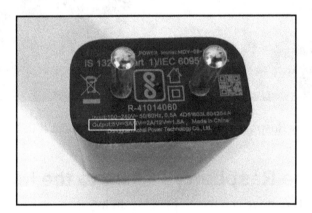

- **Power bank or battery bank**: Alternatively, we can use a power bank or a battery bank. As mentioned earlier, we need to connect the power bank to the Raspberry Pi via micro USB B port, and we also need to make sure that it provides 5V of voltage output and around 2.5A of current output:

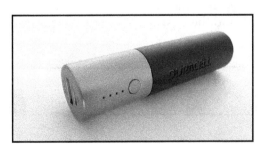

5. Once everything is plugged in, turn the display unit on and make sure that you have selected the correct HDMI option.

6. Next, turn on the power supply. You will see the red LED on your Raspberry Pi turn on. Wait for around 10-20 seconds for the Raspberry Pi to boot up. Once this is done, you will see the following screen:

Now that we've got our Raspberry Pi 3B+ running, let's connect it to the internet.

Connecting the Raspberry Pi 3B+ to the internet

There are two methods that we can use to provide an internet connection to the Raspberry Pi:

- **An Ethernet cable**: The Raspberry Pi 3B+ contains an Ethernet port. To provide an internet connection via an Ethernet port, simply connect an Ethernet cable to it.

- **Wi-Fi**: Connecting the Raspberry Pi over Wi-Fi is also pretty straightforward. Click on the Wi-Fi icon in the taskbar. Select your Wi-Fi network, enter the correct password, and the Raspberry Pi will connect to the desired Wi-Fi network:

After setting up the Raspberry Pi 3B+ as a desktop computer, we can simply open any code editor and start writing programs to control motors or LEDs with the Raspberry Pi.

Since we are going to create a movable robot using the Raspberry Pi, however, the desktop computer setup will not work. This is because the display, the keyboard, and the mouse, all of which are attached directly to the Pi, will limit its movement. In the next section, to be able to use it without these peripherals, we'll look at how to connect the Raspberry Pi 3B+ wirelessly to a laptop via Wi-Fi.

Connecting the Raspberry Pi 3B+ to a laptop via Wi-Fi

To connect the Raspberry Pi 3B+ wirelessly to a laptop via Wi-Fi, we first need to connect the RPi to a Wi-Fi network with a piece of software called PuTTY. After that, we can find out the IP address of the Raspberry Pi and enter that into a piece of software called **VNC Viewer** to connect the Raspberry Pi to a laptop. In order to perform this task successfully, the Raspberry Pi and laptop must be connected to the same Wi-Fi network.

The hardware required includes the following:

- **An Ethernet cable**: The Ethernet cable will be attached directly to the Ethernet port of the Raspberry Pi 3B+ and the Ethernet port of the laptop. If your laptop does not contain an Ethernet port, you will need to purchase a **USB-to-Ethernet** connector for your laptop.
- **Micro USB B wire**: This is a standard Micro USB B wire for connecting the Raspberry Pi 3B+ to the laptop.

The software required is **PuTTY,** VNC Viewer, and Bonjour.

Creating an SSH file on a microSD card

After installing the aforementioned software, we need an SSH file on the microSD card to enable SSH for the Raspberry Pi 3B+. To do this, perform the following steps:

1. Open the drive allotted to the SD card. In our case, this is the boot (F:) drive. As shown in the following screenshot, there are some files on the microSD card:

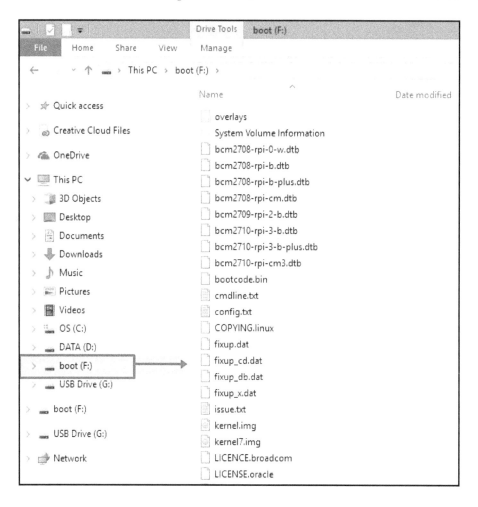

2. To create the SSH file, right-click in the drive, click on **New,** and select a **Text Document**, as shown here:

3. Give this text file the name `ssh` but do not include the `.txt` extension. We will get a warning stating that this file will become unstable because it doesn't have an extension. Click on the **Yes** button:

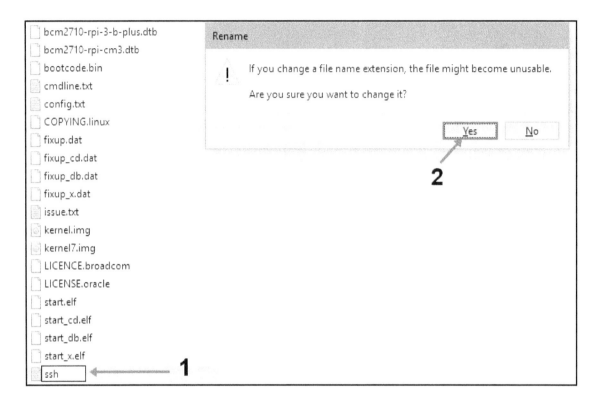

4. Next, right-click on the ssh file and select the **Properties** option. Inside **Properties**, click on the **General** tab. We should see that the **Type of file** is set to **File**. Click **OK**:

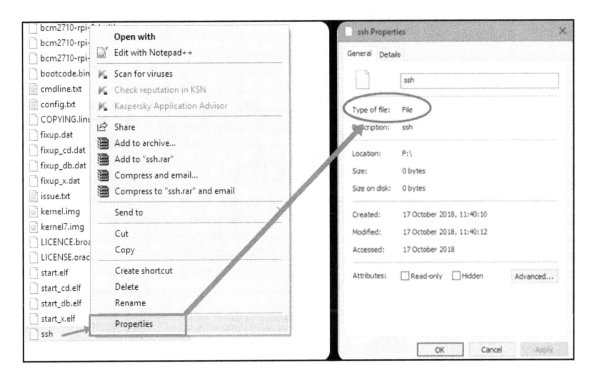

After creating an SSH file on a microSD card, remove the card reader from the laptop and insert the microSD card into the Raspberry Pi 3B+.

In the next section, we will look at how to connect the RPi 3B+ to a Wi-Fi network. The setup is done on a Windows system. If you have a Mac, then you can follow one of the following tutorial videos:

- **Access Raspbian OS on a Mac**: https://www.youtube.com/watch?v=-v88m-HYeys
- **Access Raspberry display on a VNC Viewer**: https://www.youtube.com/watch?v=PYunvpwSwGY

Connecting the Raspberry Pi 3B+ to a Wi-Fi network using PuTTY

After inserting the microSD card into the RPi, let's see how to connect the Raspberry Pi to a Wi-Fi network using PuTTY:

1. First, connect one end of the Ethernet cable to the Ethernet port of the Raspberry Pi, and the other end to the Ethernet port of the laptop.
2. Next, power up the Raspberry Pi 3B+ by connecting it to the laptop using a micro USB B cable. We will see the red power LED turn on. We will also see that the yellow LED of the Ethernet port turns on and blinks continuously.
3. After that, open the PuTTY software. Inside the **Host Name** box, type `raspberrypi.local` and click on the **Open** button:

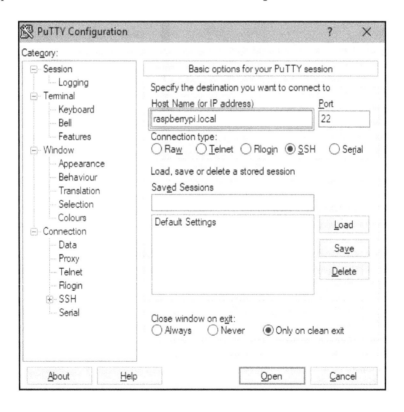

4. We will then see a **PuTTY Security Alert** message. Click **Yes**:

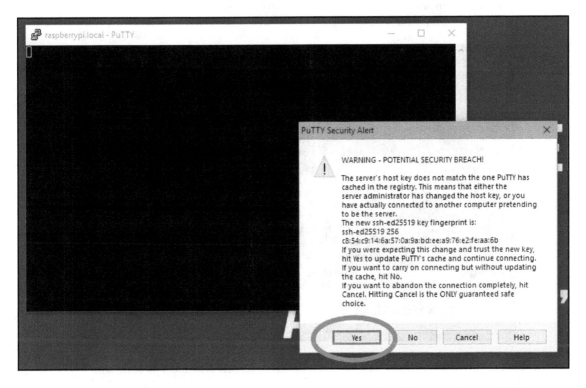

5. Inside PuTTY, we will need to enter the credentials of the Raspberry Pi. The default login is pi and the password is raspberry. After entering the password, press *Enter*:

6. After that, to connect the Raspberry Pi 3B+ to a particular Wi-Fi network, enter the `sudo nano /etc/wpa_supplicant/wpa_supplicant.conf` command as shown in this screenshot:

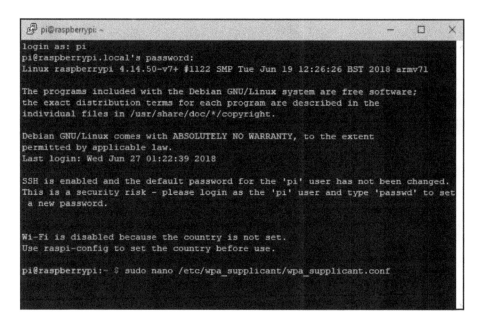

7. This command will open the nano editor, which will look as follows:

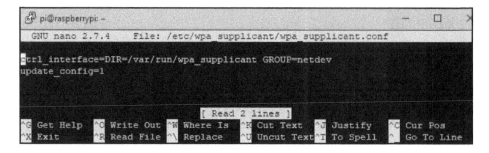

8. Below the `update_config=1` line, enter the name and password for your Wi-Fi, following this syntax:

```
network={
ssid="Wifi name"
psk="Wifi password"
}
```

Make sure that you add the preceding code exactly below the `update_config=1` line. The Wi-Fi name and the password should be in double quotes (`""`), as shown here:

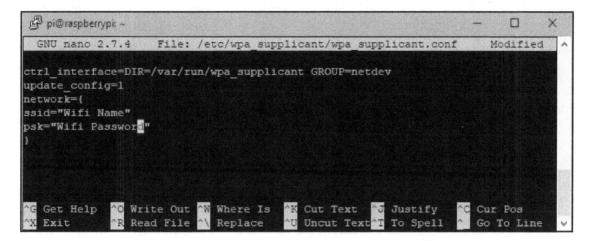

After entering the Wi-Fi name and password, press the *Ctrl* + *O* keys to save the changes. Then, press *Enter*. After that, press the *Ctrl* + *X* keys to exit the nano editor.

9. To reconfigure and connect the Raspberry Pi to the Wi-Fi network, enter the following command: `sudo wpa_cli reconfigure`. If you have connected successfully, you will see the type of the interface and an `OK` message:

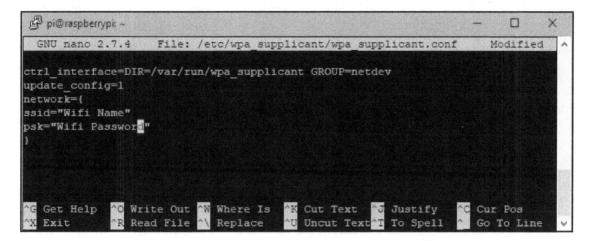

10. We will then need to restart the Raspberry Pi. To do this, type `sudo shutdown now`. Once the Raspberry Pi is shut down, close the PuTTY software.

11. Next, unplug the USB cable from the laptop.

12. After that, unplug the Ethernet cable that is connected to the Raspberry Pi and the laptop.
13. Re-connect the USB cable so that the Raspberry Pi turns on.
14. Open PuTTY. Inside the **Host Name** field, enter `raspberrypi.local` again and press the **Open** button:

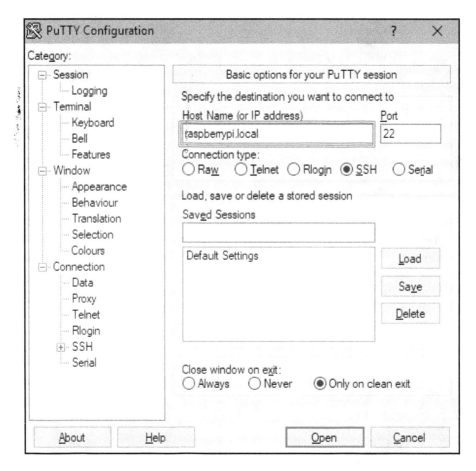

15. Enter the username and password that we used previously.

16. Once the Raspberry Pi is connected to the Wi-Fi network, the Wi-Fi network will assign it an IP address. To find the IP address, enter the `ifconfig wlan0` command and press *Enter*. You will notice that an IP address has now been assigned:

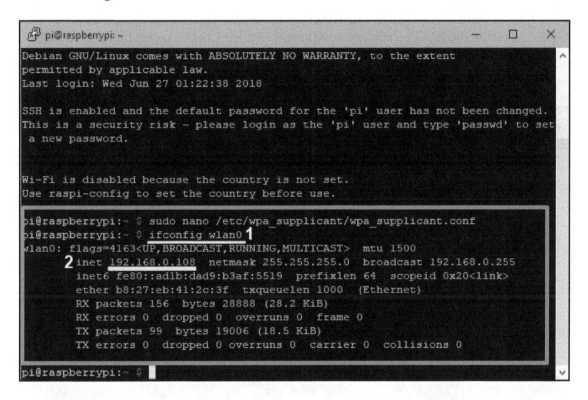

In my case, the IP address is `192.168.0.108`. Please note down your IP address somewhere, as you will need to enter it when using the VNC Viewer software.

Enabling the VNC server

To view the Raspberry Pi display, we will need to enable the VNC server from the Raspberry Pi configurations window:

1. To open the configurations window, we need to type `sudo raspi-config` inside the PuTTY Terminal and press *Enter*. We can then open the **Interfacing Options** as shown here:

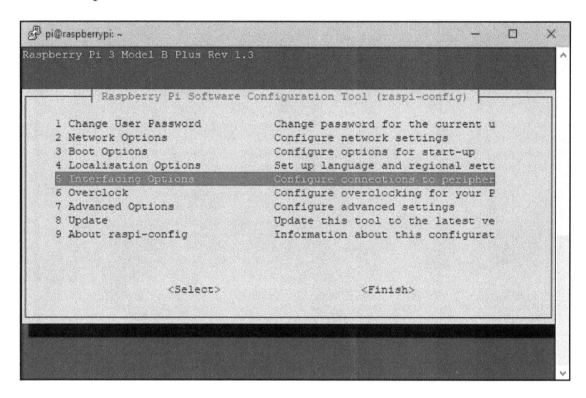

2. We can then open the **VNC** options:

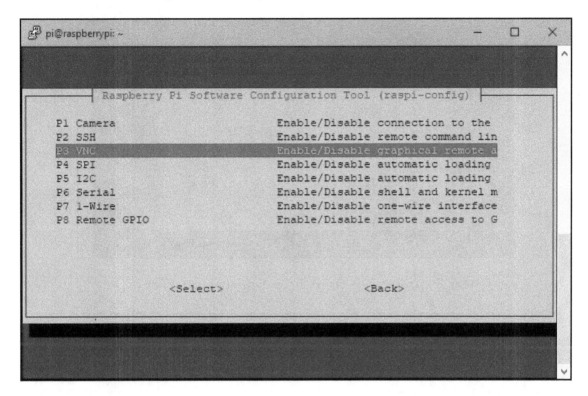

3. To enable the VNC server, navigate to the **Yes** option and press *Enter:*

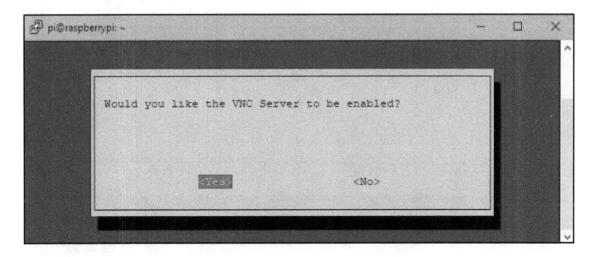

4. After enabling the VNC server, press **Ok**:

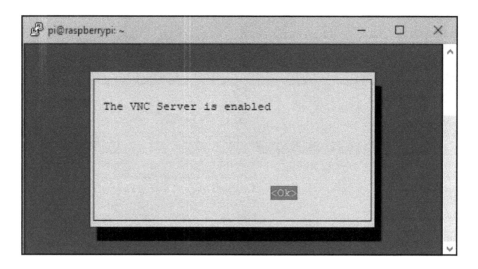

5. Press **Finish** to exit the Raspberry Pi configuration window:

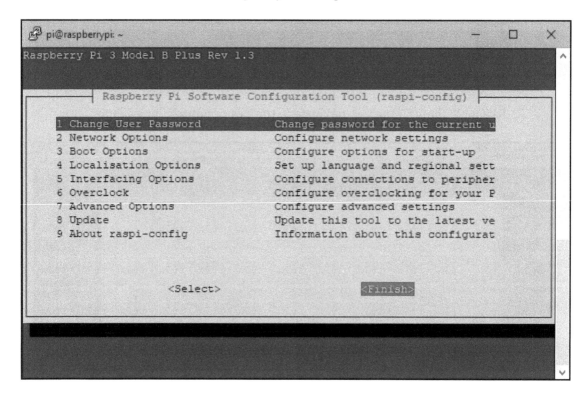

After enabling the VNC server, we will open the VNC Viewer software so that we can see the Raspberry Pi display.

Viewing the Raspberry Pi output on the VNC Viewer

To view the Raspberry Pi output on the VNC viewer, following the instructions below:

1. After opening the VNC Viewer software, enter the IP address of your Raspberry Pi inside the VNC Viewer and press *Enter*:

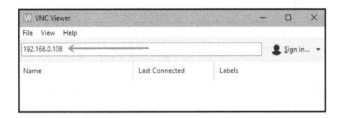

2. You will get a pop-up message stating that VNC Viewer has no record of this VNC server. Press **Continue**:

3. Enter the username as `pi` and the password as `raspberry`. Select the **Remember password** option and press **OK**:

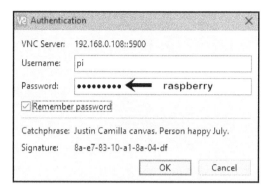

We should now be able to view the Raspberry Pi display output inside the VNC Viewer software:

Now that we have connected the Raspberry Pi to the laptop via Wi-Fi, there is no need to connect the Raspberry Pi to the laptop via a USB cable. Next time, we can simply power the Raspberry Pi using a power bank or mobile charger. When we select the IP address of our Raspberry Pi, we can view the Raspberry Pi display output using the VNC Viewer software.

As mentioned already, please make sure that both the Raspberry Pi and the laptop are connected to the same Wi-Fi network when using the laptop for remote desktop access.

Increase the VNC's screen resolution

After viewing the RPi's display output in the VNC Viewer, you will notice that the screen resolution of the VNC Viewer is small and it does not cover the entire screen. To increase the screen resolution, we need to edit the `config.txt` file:

1. Enter the following command in the terminal window:

    sudo nano /boot/config.txt

2. Next, below the `#hdmi_mode=1` code, enter the following three lines:

    ```
    hdmi_ignore_edid=0xa5000080
    hdmi_group=2
    hdmi_mode=85
    ```

3. After this, press *Ctrl + O* and then press *Enter* to save the file. Press *Ctrl + X* to exit:

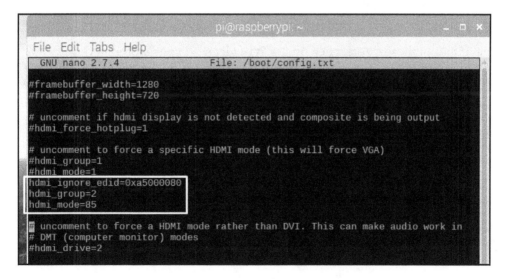

4. Next, reboot your RPi to apply these changes:

    sudo reboot

Once rebooted, you will notice that the VNC's screen resolution has increased and it now covers the entire screen.

Handling VNC and PuTTY errors

In the VNC Viewer, sometimes, when you are selecting the IP address of the RPi, you may see the following pop-up error message instead of the RPi display:

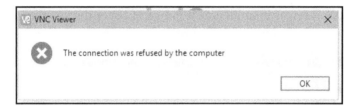

You may also see the following message:

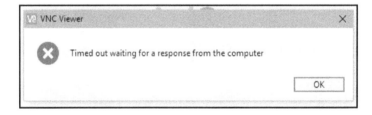

If you get either of these errors, click on the Wi-Fi icon on your laptop and make sure that you are connected to the same Wi-Fi network to which the RPi is connected. If this is the case, there is a chance that the IP address of your RPi has been changed inside the Wi-Fi network, which sometimes happens when a new device is connected to the Wi-Fi network.

To find the new IP address follow these steps:

1. Open PuTTY and type `raspberrypi.local` inside the **Hostname** box.
2. Enter the command `ifconfig wlan0` inside PuTTY's Terminal window. If your IP address has changed, you will notice the new IP address in the `inet` option.
3. Enter the new IP address inside VNC Viewer to view the RPi display output.

Sometimes, you may be unable to connect to Putty as well, and you will see the following error:

To solve the preceding error in PuTTY, follow these steps:

1. Connect a LAN cable to the RPi and the laptop.
2. Power on your RPi and try connecting to putty by entering `raspberrypi.local` in the hostname box. With the LAN cable connected to the RPi and the laptop, you should be able to access the PuTTY Terminal window.
3. Follow the previous steps to find the RPi's new IP address.
4. Once you see the RPi display inside VNC Viewer, you can remove the LAN cable.

Setting up the Raspberry Pi Zero W as a desktop computer

As we've said, the Raspberry Pi Zero W is a stripped-down version of the Raspberry Pi 3B+. The Raspberry Pi Zero W has very limited connections, so in order to connect it to different peripherals, we will need to purchase some additional components. We will need the following hardware components:

- A keyboard
- A mouse
- A microSD card of minimum 8 GB (recommended 32 GB)
- A microSD card reader
- An HDMI cable
- A display unit, preferably an LED screen or a TV with an HDMI port
- A mobile charger or a power bank to power the Raspberry Pi

- A micro USB B-to-USB connector (also known as an OTG connector), which looks like the following:

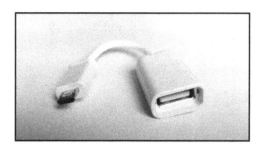

- A mini HDMI-to-HDMI connector, as follows:

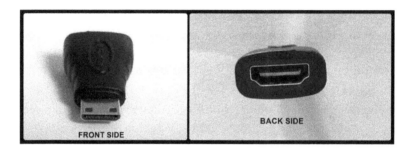

- A USB hub, as pictured here:

Now that we know what hardware we require, let's set up our Raspberry Pi Zero W.

Setting up the Raspberry Pi Zero W

The steps to install the Raspbian OS onto a microSD card are exactly same as those listed already for the **Raspberry Pi 3B+** in the section *Installing Raspbian OS on an SD card*. Once you have your SD card ready, follow these steps to set up the Raspberry Pi Zero W:

1. First, insert the microSD card into the SD card slot of the Raspberry Pi Zero W.
2. Connect one end of the **mini HDMI to the HDMI connector** (**H2HC**) inside the HDMI port of the Raspberry Pi Zero W, and the other end of H2HC connector to the HDMI cable.
3. Connect the OTG connector to the Micro USB data port—not the power port—and connect the USB hub to the OTG connector.
4. Connect the keyboard and the mouse to the USB hub.
5. Connect a 5V mobile charger or a battery bank to the power unit's Micro USB port.
6. Next, connect an HDMI cable to the HDMI port of a TV or a monitor.
7. Connect the mobile charger to your mains supply to power the Raspberry Pi Zero W. You will then see the green LED blink for a period of time as the Raspberry Pi Zero W turns on.
8. If you have connected the HDMI cable to your TV, select the correct HDMI input channel. The following annotated photo shows the connections mentioned here:

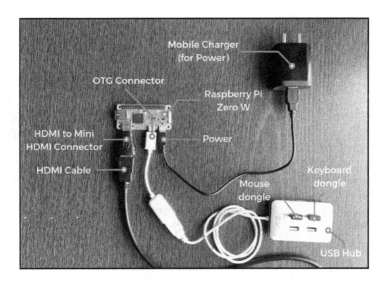

Raspberry Pi Zero W Connections

9. It will take around two or three minutes for the Raspberry Pi Zero W to boot up. Once it's ready, you will see the following window on your TV or monitor screen:

10. To shut down the Raspberry Pi, press the Raspberry Pi icon and click on **Shutdown**.

Now that it's set up, let's connect our Raspberry Pi Zero W to a laptop.

Connecting the Raspberry Pi Zero W to a laptop via Wi-Fi

When the Raspberry Pi Zero first came out in 2015, it didn't have a built-in Wi-Fi module, which made it difficult to connect to the internet. Some Raspberry Pi developers came up with useful hacks to connect the Raspberry Pi to the internet and some companies also created Ethernet and Wi-Fi modules for the Raspberry Pi Zero.

In 2017, however, the Raspberry Pi Zero W was launched. This had a built-in Wi-Fi module, which meant that Raspberry Pi developers no longer needed to perform any DIY hacks or purchase a separate component to add internet connectivity. Having built-in Wi-Fi also helps us to wirelessly connect the Raspberry Pi Zero W to a laptop. Let's take a look at how this can be done.

The process of connecting Raspberry Pi Zero W to a laptop's Wi-Fi is similar to that of Raspberry Pi 3B+. Since the Raspberry Pi Zero W does not have an Ethernet port, however, we will have to write a few lines of code inside the `cmdline.txt` and `config.txt` files.

Even though `cmdline.txt` and `config.txt` are **text** (**TXT**) files, the code in these files does not open properly inside Microsoft's Notepad software. To edit these files, we will need to use code editor software, such as Notepad++ (only available for Windows) or Brackets (available for Linux and macOS).

After installing either of these, let's customize the microSD card as follows:

1. In the Raspberry Pi Zero W, we also need to create an SSH file on the microSD card. For instructions on how to create an SSH file on the microSD card, refer to the section, *Creating an SSH file on a microSD card*.
2. After creating an SSH file, right-click on the `config.txt` file and open it in Notepad++ or Brackets. In this case, we will open it in Notepad++:

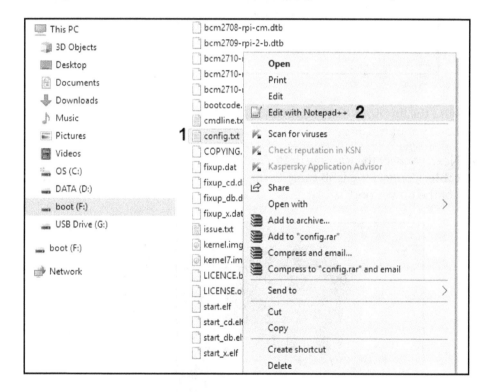

Scroll all the way down to the bottom of this code and add the line `dtoverlay=dwc2` at the end. After adding the code, save and close the file.

3. Next, open the `cmdline.txt` file inside Notepad++. The entire code inside the `cmdline` file will be displayed on one line. Next, make sure that you add only one space between the word `consoles` and the word `modules`.

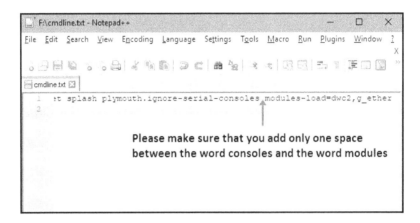

Enter the line `modules-load=dwc2,g_ether` at the end next to the `plymouth.ignore-serial-consoles` code:

4. Next, connect the Raspberry Pi Zero W to your laptop using a **data transfer USB cable**. Connect the USB cable to the data port of the Raspberry Pi Zero W, and not the power port:

5. Make sure that the USB cable that you are connecting to the Raspberry Pi Zero W and the laptop supports data transfer. For example, take a look at the following photograph:

In the preceding photo, there are two similar, but importantly different, cables:

- The small USB cable on the left was included with my power bank kit. This USB cable supplies power but does not support data transfer.
- The USB cables on the right were included with the purchase of a new Android smartphone. These do support data transfer.

A simple way to check whether your USB supports data transfer or not is to connect it to your smartphone and laptop. If your smartphone is detected, this means that your USB cable does support data transfer. If not, you will need to purchase a USB cable that supports data transfer. The following screenshot shows a smartphone being detected by a PC, meaning that the cable in use is a data cable:

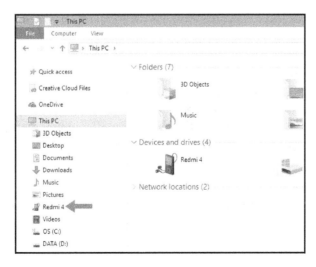

If your USB cable is detected but loses connection frequently, I recommend that you purchase a new USB cable. Sometimes, older USB cables do not work properly because of wear and tear.

Connecting the Raspberry Pi Zero W to a Wi-Fi network using PuTTY

To connect the Raspberry Pi Zero W to a Wi-Fi network, refer to the *Connecting the Raspberry Pi 3B+ to a Wi-Fi network using PuTTY* section. The steps for connecting a Raspberry Pi Zero W to a Wi-Fi network are exactly the same.

Enabling VNC Viewer for Raspberry Pi Zero W

To enable a VNC Viewer for the Raspberry Pi Zero W, refer to the *Enabling the VNC server* section.

Viewing Raspberry Pi Zero W output on VNC Viewer

To view the Raspberry Pi Zero W output in the VNC Viewer, refer to the *Viewing the Raspberry Pi output on VNC Viewer* section.

Summary

In this chapter, we've looked at how to set up the Raspberry Pi 3B+ and the Raspberry Pi Zero W as a normal desktop computer. We also learned how to connect a Raspberry Pi to a laptop wirelessly via a Wi-Fi network. You can now control a Raspberry Pi remotely from your laptop without needing to connect the Raspberry Pi to a keyboard, mouse, and monitor.

In the next chapter, we will first look at some basic commands for maneuvering around the Raspberry Pi OS. We will install a C++ library called Wiring Pi onto our Raspberry Pi, and gain an understanding of the pin configurations for this library. Finally, we will write our first C++ program and upload it wirelessly to our Raspberry Pi.

Questions

1. Which processor is present on the Raspberry Pi 3B+?

2. How many GPIO pins are present on the Raspberry Pi 3B+?

3. Which software are we using for viewing the Raspberry Pi display output on our laptop?

4. What is the default username and password of a Raspberry Pi?

5. What is the command for accessing the configurations inside the Raspberry Pi?

2
Implementing Blink with wiringPi

After setting up the Raspberry Pi, it's now time to connect different electronics components to it and program it using the C++ programming language. To use C++, we will first have to download and install a library called **wiringPi**.

In this chapter, we will cover the following topics:

- Installing the `wiringPi` library inside the Raspberry Pi
- Making an LED blink
- Smart Light—working with a digital sensor
- Pulse Width Modulation using softPwm

Technical requirements

The hardware requirements for this chapter are as follows:

- 1 LED (any color)
- 1 **LDR** (**Light Dependent Resistor**) sensor module
- Raspberry Pi 3B+
- 5-6 female to female connecting wires

The code files for this chapter can be downloaded from `https://github.com/ PacktPublishing/Hands-On-Robotics-Programming-with-Cpp/tree/master/Chapter02`.

Installing the wiringPi library in the Raspberry Pi

wiringPi is a pin-based GPIO access library that is written in C. Using this library, you can control the Raspberry Pi using C/C++ programming. The `wiringPi` library is easy to set up. Once installed, the Raspberry Pi GPIO pins will have wiringPi pin numbering. Let's take a look at how to download and install wiringPi:

1. First, open the Terminal window by clicking on its icon from the taskbar:

2. Before installing the `wiringPi` library, we first need to verify that our Raspberry Pi is up to date by checking for updates. If your Raspberry Pi is not updated, you may face errors while installing the `wiringPi` library. To update your Raspberry Pi, type the following command:

```
$ sudo apt-get update
```

The output of the preceding command can be seen as follows:

```
                              pi@raspberrypi ~                          _  □  ✕
File  Edit  Tabs  Help
pi@raspberrypi:~ $ sudo apt-get update
Get:1 http://raspbian.raspberrypi.org/raspbian stretch InRelease [15.0 kB]
Get:2 http://archive.raspberrypi.org/debian stretch InRelease [25.3 kB]
Get:3 http://raspbian.raspberrypi.org/raspbian stretch/main armhf Packages [11.7
 MB]
Get:4 http://archive.raspberrypi.org/debian stretch/main armhf Packages [181 kB]
Get:5 http://archive.raspberrypi.org/debian stretch/ui armhf Packages [34.3 kB]
30% [3 Packages 1,562 kB/11.7 MB 13%]                    52.0 kB/s 3min 17s
```

 Depending on your internet speed, it will take around 10-15 minutes for the updates to download and install. Make sure that you place your Raspberry Pi near your Wi-Fi router.

3. After the update, type in the following command to upgrade the Raspberry Pi:

    ```
    $ sudo apt-get upgrade
    ```

    While upgrading, you may get a message asking you to download a particular component. Type Y and then press *Enter*. It will take around 30-40 minutes for the upgrade to complete. Once the upgrade is done, you will see the following message:

```
Updating certificates in /etc/ssl/certs...
0 added, 0 removed; done.
Running hooks in /etc/ca-certificates/update.d...
done.
pi@raspberrypi:~ $
```

4. After updating your Raspberry Pi, you will need to download and `install` `git-core` inside your Raspberry Pi. To install Git, type the following command:

```
$ sudo apt-get install git-core
```

```
Processing triggers for libc-bin (2.24-11+deb9u3) ...
Processing triggers for ca-certificates (20161130+nmu1+deb9u1) ...
Updating certificates in /etc/ssl/certs...
0 added, 0 removed; done.
Running hooks in /etc/ca-certificates/update.d...
done.
pi@raspberrypi:~ $ sudo apt-get install git-core  ←
Reading package lists... Done
Building dependency tree
Reading state information... Done
The following NEW packages will be installed:
  git-core
0 upgraded, 1 newly installed, 0 to remove and 2 not upgraded.
Need to get 1,414 B of archives.
After this operation, 8,192 B of additional disk space will be used.
Get:1 http://raspbian.raspberrypi.org/raspbian stretch/main armhf git-core all 1
:2.11.0-3+deb9u4 [1,414 B]
Fetched 1,414 B in 0s (2,276 B/s)
Selecting previously unselected package git-core.
(Reading database ... 115961 files and directories currently installed.)
Preparing to unpack .../git-core_1%3a2.11.0-3+deb9u4_all.deb ...
Unpacking git-core (1:2.11.0-3+deb9u4) ...
Setting up git-core (1:2.11.0-3+deb9u4) ...
pi@raspberrypi:~ $
```

5. After this, to download the `wiringPi` library from `git`, type in the following command:

```
git clone git://git.drogon.net/wiringPi
```

```
pi@raspberrypi:~ $ git clone git://git.drogon.net/wiringPi
Cloning into 'wiringPi'...
remote: Counting objects: 1177, done.
remote: Compressing objects: 100% (980/980), done.
remote: Total 1177 (delta 822), reused 212 (delta 142)
Receiving objects: 100% (1177/1177), 369.48 KiB | 247.00 KiB/s, done.
Resolving deltas: 100% (822/822), done.
pi@raspberrypi:~ $
```

6. Now, if you click on the **File Manager** option and click on the pi folder, you should see the wiringPi folder:

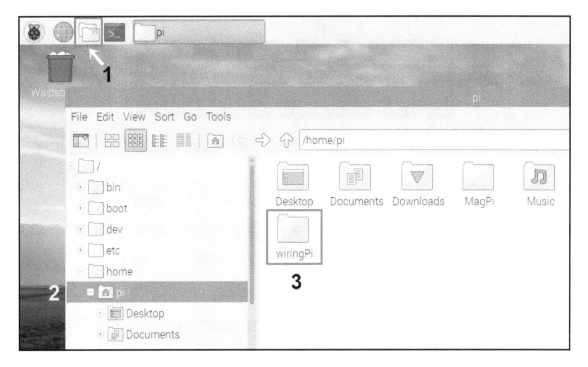

7. Next, change the directory to wiringPi, so that the wiringPi files are downloaded and installed inside this particular folder. The command for changing the directory is cd:

> **$ cd ~/wiringPi** (The ~ symbol is above the Tab key and it points to pi directory)

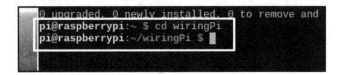

You should now see the directory pointing toward the wiringPi folder.

8. Next, to fetch the Git files from their `origin` directory, type in the following command:

```
$ git pull origin
```

9. Finally, for building the files, type in the following command:

```
$ ./build
```

Once everything is done, you will see an `All done` message:

Now that we have installed the wiringPi library, we can move on and understand wiringPi pin configurations on the RPi.

Accessing Raspberry Pi GPIO pins via wiringPi

Since we have installed wiringPi, we can now look at the wiringPi pin numbering, as shown in the following screenshot:

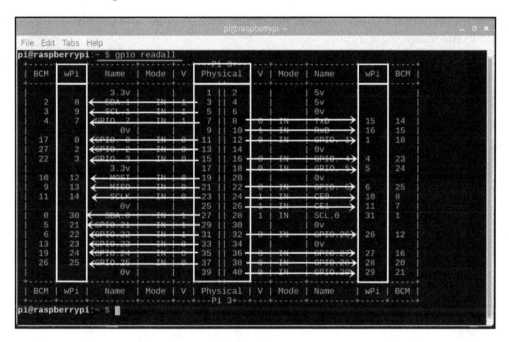

The `Physical` column represents the Raspberry Pi numbering from 1-40. On both sides of the `Physical` column, you will see the wiringPi (`wPi`) columns. The arrows pointing from the `Physical` column to `wPi` represent the wiringPi pin numbering for a particular physical pin of a Raspberry Pi.

Take a look at the following examples:

- Physical pin number 3 has a wiringPi pin number of 8
- Physical pin number 5 has a wiringPi pin number of 9
- Physical pin number 8 has a wiringPi pin number of 15
- Physical pin number 11 has a wiringPi pin number of 0
- Physical pin number 40 has a wiringPi pin number of 29

By consulting this table, you can figure out which of the remaining physical pins correspond to which wiringPi pins.

wiringPi pin numbers from **17-20** do not exist. After **wPi pin 16**, we skip straight to **wPi pin 21**.

To better understand the relationship between the wiringPi pins and the physical pins, you can refer to the following diagram:

Raspberry Pi Pinout

3v3 Power		1	2	5v Power	
BCM 2	(WiringPi 8)	3	4	5v Power	
BCM 3	(WiringPi 9)	5	6	Ground	
BCM 4	(WiringPi 7)	7	8	BCM 14	(WiringPi 15)
Ground		9	10	BCM 15	(WiringPi 16)
BCM 17	(WiringPi 0)	11	12	BCM 18	(WiringPi 1)
BCM 27	(WiringPi 2)	13	14	Ground	
BCM 22	(WiringPi 3)	15	16	BCM 23	(WiringPi 4)
3v3 Power		17	18	BCM 24	(WiringPi 5)
BCM 10	(WiringPi 12)	19	20	Ground	
BCM 09	(WiringPi 13)	21	22	BCM 25	(WiringPi 06)
BCM 11	(WiringPi 14)	23	24	BCM 8	(WiringPi 10)
Ground		25	26	BCM 7	(WiringPi 11)
BCM 0	(WiringPi 30)	27	28	BCM 1	(WiringPi 31)
BCM 5	(WiringPi 21)	29	30	Ground	
BCM 6	(WiringPi 22)	31	32	BCM 12	(WiringPi 26)
BCM 13	(WiringPi 23)	33	34	Ground	
BCM 19	(WiringPi 24)	35	36	BCM 16	(WiringPi 27)
BCM 19	(WiringPi 25)	37	38	BCM 20	(WiringPi 28)
Ground		39	40	BCM 21	(WiringPi 29)

The wiringPi pin numbering is what you will need to remember while programming. We can use a total of **28** wiringPi pins for programming. As well as these, we have the following pins, which can be used for providing power and can be used as ground pins:

- Physical pin numbers **6, 9, 14, 20, 25, 30, 34,** and **39** are ground pins
- Physical pin numbers **2** and **4** provide a +5V supply
- Physical pin numbers **1** and **17** provide a +3.3V supply

Let's move on to writing our first C++ program for Raspberry Pi.

Making an LED blink

The very first project that we are going to create is making an LED blink. For this project, we require the following hardware components:

- Raspberry Pi
- 1 LED
- Two female-to-female wires

Wiring connections

Connecting the LED to the Raspberry Pi is straightforward. Before doing this, however, let's take a closer look at the pins of the LED:

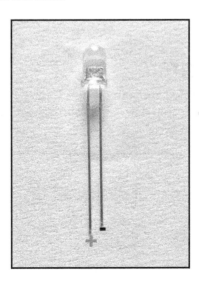

The LED contains one positive pin and one negative pin. The long pin is the positive pin, which you can connect to any data pin of the Raspberry Pi. The short pin is the negative pin, which can be connected to the ground pin of the Raspberry Pi.

Let's connect it up. First, connect the negative pin of the LED to the ground pin (physical pin number **6**) of the Raspberry Pi. Next, connect the positive pin of the LED to wiringPi pin number **15**:

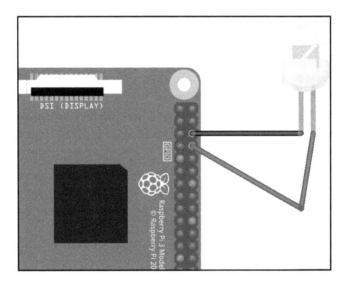

Now the we've connected the LED to the Raspberry Pi, let's write a program to make the LED blink.

The blinking program

To write our first C++ program, we are going to use **Geany Programmer's Editor**. To open Geany, click on the **Raspberry icon**, go to **Programming**, and then select **Geany Programmer's Editor**:

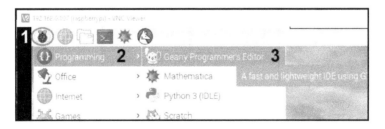

After opening Geany, you will see an unsaved file called Untitled. The first thing that we need to do is save the file. Click on **File** | **Save as** and give this file the name Blink.cpp.

Inside this file, write the following code to make the LED blink. You can download the Blink.cpp program from the Chapter02 folder of GitHub repository:

```
#include <iostream>
#include <wiringPi.h>

int main(void)
{
wiringPiSetup();
pinMode(15,OUTPUT);

  for(;;)
  {
digitalWrite(15,HIGH);
delay(1000);
digitalWrite(15,LOW);
delay(1000);
  }
return 0;
  }
```

If you have done Arduino programming before, you are likely to have understood around 90% of this code. This is because wiringPi allows us to write C++ programs in Arduino format:

1. In the preceding code, we first import the iostream and the wiringPi library.
2. Next, we have the main function, called int main. Since this function does not have any arguments, we write a void statement inside the round brackets.
3. After this, the wiringPisetup() function initializes wiringPi. It assumes that this program will use the wiringPi numbering scheme.
4. Next, with the pinMode(15, OUTPUT) command, we are setting the wiringPi pin number 15 as the OUTPUT pin. This is the pin we have connected to the positive pin of the LED.
5. After that, we have an infinite for loop. The code written inside it will run infinitely, unless we stop it manually from the coding editor.
6. With the digitalWrite(15,HIGH) command, we write a HIGH signal on the LED, which means the LED will turn on. Instead of HIGH, we could also put the number 1.
7. After this, with the delay(1000) command, we ensure that the LED is only **on** for one second.

8. Next, with the `digitalWrite(15,LOW)` command, we write a `LOW` signal on the LED. This means that the LED will turn **off** for one second.

9. Since this code is inside a for loop, the LED will keep turning **on** and **off** until we instruct it otherwise.

Uploading the code to the Raspberry Pi

Since we are using wiringPi numbering conventions, we will add the - `lwiringPi` command inside the **Build** command so that our C++ program with the `wiringPi` library is compiled and built successfully. To open the **Build** command, click on **Build** | **Set Build Commands**. Inside the command boxes next to the **Compile** and **Build** buttons, add `-lwiringPi` at the end and then click **OK**:

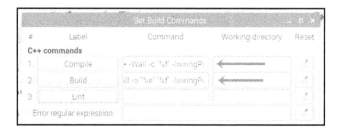

Next, to compile the code, click on the **compilation button** (the brown icon). Finally, to upload the code to the Raspberry Pi, press the **build button** (the airplane icon):

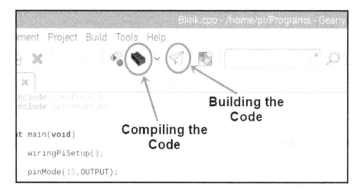

The compilation icon will check for errors in the code. If there aren't any, click the build icon to test the blinking output. After building the code, the build icon will turn into a red circle. Click on the red circle to stop the program.

Smart light – working with digital sensor

After writing our first C/C++ program for Raspberry Pi, we can now write a program that will take input from an LDR sensor and turn the LED on or off. For this project, you will need the following hardware components:

- 1 LDR sensor module
- 1 LED
- Raspberry Pi
- 5 female-to-female connecting wires

First, let's explore how the LDR sensor works.

The LDR sensor and the way it works

An LDR sensor is an analog input sensor that consists of a variable resistor whose resistance varies depending on the amount of light falling on its surface. When there is no light in the room, the resistance of the LDR sensor is HIGH (up to 1 M ohm) and in the presence of light, the resistance of the LDR sensor is LOW. The LDR sensor consists of two pins. These pins do not have positive and negative polarity. We can use any pin as a data or ground pin and because of this, the LDR sensor is sometimes referred to as a special type resistor. The image of the LDR sensor is shown in the following photo:

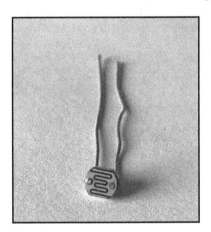

Since LDR is an analog sensor, we cannot connect it directly to the RPi as this does not contain an **analog to digital converter** (**ADC**) circuit. Because of this, RPi cannot read incoming analog data from an LDR sensor. So, instead of an LDR sensor, we will use an LDR digital sensor module, which will provide digital data to the RPi:

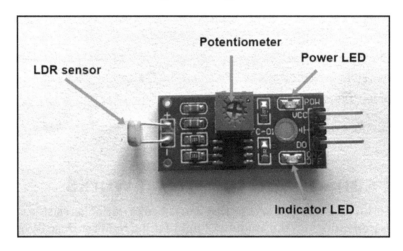

An LDR sensor module will read the incoming analog data from the LDR sensor and provide digital data in the form of HIGH or LOW as output. The LDR sensor module consists of 3 pins: **D0** (**data output**), ground, and Vcc. D0 will provide digital data as output, which is further provided as input to RPi pins. The D0 pin will be HIGH in low light and will be LOW in the presence of light. The sensor module also consists of a potentiometer sensor, which can be used to vary the resistance of the LDR sensor.

Practical uses of the LDR sensor module is seen in street lamps, which automatically turn off during day time and turn on during night time. The smart light program that we are going to write is somewhat similar to this application, but instead of a street lamp, we are going to use an LED to keep things simple.

Now that we've gained an understanding of the basic way in which an LDR sensor works, next let's connect the LDR sensor module to Raspberry Pi.

Wiring connection

With a wiring connection, we can connect the LDR sensor module and a LED to the RPi:

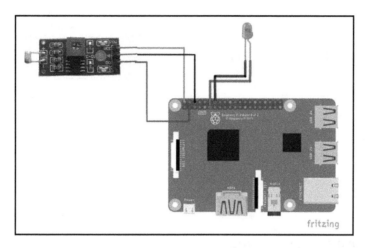

The wiring connections are as follows:

- wiringPi pin 8 of the RPi is connected to the D0 pin of the LDR sensor module
- Physical pin 2 of the RPi is connected to the Vcc pin of the LDR sensor module
- Physical pin 6 of the RPi is connected to the Gnd pin of the LDR sensor module
- wiringPi pin 0 is connected to the positive pin of the LED
- Physical pin 14 is connected to the negative pin of the LED

Now that we have connected the LDR sensor module and LED to the RPi, let's write the program to turn the LED on/off by taking inputs from the LDR sensor.

Smart light program

In this smart light program, we will first read input from the LDR sensor and, based on the input value, we will turn the LED on or off. The program for smart light is described as follows. You can download the SmartLight.cpp program from the Chapter02 folder of this book's GitHub repository:

```
#include <iostream>
#include <wiringPi.h>

int main(void)
{
```

```
wiringPiSetup();

pinMode(0,OUTPUT);
pinMode(8,INPUT);

for(;;)
{
int ldrstate = digitalRead(8);
if(ldrstate == HIGH)
{
digitalWrite(0,HIGH);
}
else
{
digitalWrite(0,LOW);
}
}
return 0;
}
```

The explanation of the preceding program is as follows:

- Inside the `main` function, we have set wiringPi pin 8 as the input pin and wiringPi pin 0 as the output pin.
- Next, in the `for` loop, using the `digitalRead(8)` function, we are reading the incoming digital data from the digital pin(D0) of the LDR sensor and storing it inside the `ldrstate` variable. From the LDR sensor, we will receive HIGH(1) data or LOW(0) data. The `ldrstate` variable will be HIGH when there is no light and it will be LOW when there is light.
- Next, we will check whether the data inside the `ldrstate` variable is HIGH or LOW using an `if...else` condition.
- Using `if(ldrstate == HIGH)`, we are comparing whether the data inside the `ldrstate` variable is HIGH. If it is HIGH, we are turning the LED on using `digitalWrite(0,HIGH)`.
- If the `ldrstate` is LOW, then the `else` condition will execute and by using `digitalWrite(0,LOW)`, we are turning the LED off. Next, you can click on the **Compile** button to compile the code and then test it by clicking the **Build** button.

Now that we understand the SmartLight program, we will explore the concept of **Pulse Width Modulation** (**PWM**) and use a library called softPWM to change the brightness of an LED.

Pulse Width Modulation using softPWM

PWM is a powerful technique that can use to control the power that's delivered to an electronic component like LEDs and motors. Using PWM, we can perform operations like controlling the brightness of an LED or reducing the speed of a motor. In this section, we will first understand the way in which a PWM works and then we will write a simple PWM program to increase the brightness of an LED, step by step.

How PWM works

In the previous `Blink.cpp` program, we applied a digital signal from the RPi to the LED. Digital signals either have a HIGH state or a LOW state. In, HIGH, state the Raspberry Pi pins produces a voltage of 3.3V and in a LOW state, the pins produce a voltage of 0V. Consequently, at 3.3V, the LED is on with full brightness and at 0V, the LED is turned off:

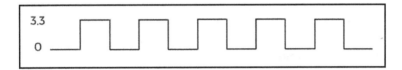

To reduce the brightness of the LED, we need to reduce the voltage. To reduce the voltage, we use PWM. In PWM, a single wave with one full repetition is called a cycle and the time taken for a cycle to complete itself is called a period. In the following diagram, the red lines represent one complete cycle. The time taken to complete that cycle is called a period:

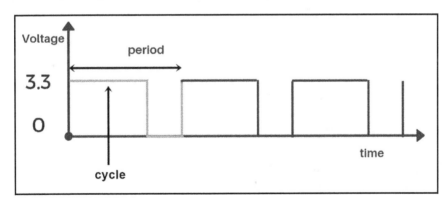

The time duration for which a signal remains HIGH is called a duty cycle, as shown in the following diagram:

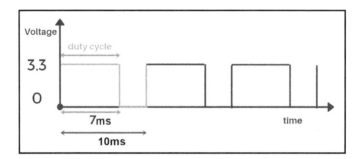

The duty cycle is represented in percentage format, and the formula for calculating the duty cycle is as follows:

Duty cycle = (time duration for HIGH signal / total time) X 100

In the preceding diagram, the signal remains HIGH for 7 milliseconds, and the total time period for a single cycle is 10 milliseconds:

$$Duty cycle = \frac{7ms}{10ms} X100$$

Duty cycle = 70% or 0.7

Consequently, the duty cycle is 0.7 or 70%. Next, to find the new voltage value, we need to multiply the duty cycle with the maximum voltage value, which is 3.3V:

Vout = Duty cycle X Vmax

Vout = 0.7 X 3.3

Vout = 2.31V

At a duty cycle of 70%, the voltage that's provided to the LED will be 2.31V and the brightness of the LED will reduce slightly.

Now, if we reduce the duty cycle to 40%, then the voltage that's provided to the LED will be 1.32V, as shown in the following diagram:

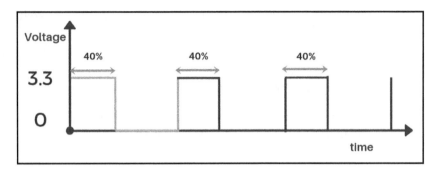

Now that we have understand how PWM is used to reduce the voltage at RPi data pins, let's take a look at the softPWM library, using which data pins can be used as PWM pins.

The softPWM library

wiringPi consists of a softPWM library, using which we can get PWM signal output from any data pin of the RPi. The softPWM library contains two main functions: softPwmCreate and softPwmWrite. Both of these functions work as follows:

```
softPwmCreate(pin number, initial duty cycle value, max duty cycle value);
```

The softPwmCreate function is used to create a PWM pin. It consists of three main parameters:

- pin number: Pin number represents the wiringPi pin that we want to set as a PWM pin.
- initial duty cycle value: In initial duty cycle value we have to provide as the minimum value of the duty cycle. The initial duty cycle value is ideally set to 0.
- max duty cycle value: In the max duty cycle value, we have to provide the maximum value of the duty cycle. This value must be set to 100:

```
softPwmWrite(pin number, duty cycle value);
```

The `softPwmWrite` function is used to write PWM data on the output device (for example, LED). It consists of two parameters:

- `pin number`: Pin number represents the wiringPi pin on which we have to write the PWM data.
- `duty cycle value`: In this parameter, we have to provide the duty cycle value. The duty cycle value must be between the initial duty cycle value and max duty cycle value, that is, in the range of 0 to 100.

Now that we understand the two functions inside the softPWM library, we will write a simple C++ program to make an LED blink at different intensities.

Making an LED blink with the softPWM library

For the blinking LED program using softPWM, you will need one led. In my case, I have connected the negative pin of the LED to physical pin 6 (ground pin) of the RPi, and the positive pin of the LED is connected to wiringPi pin 15. The wiring connection is shown in the following diagram:

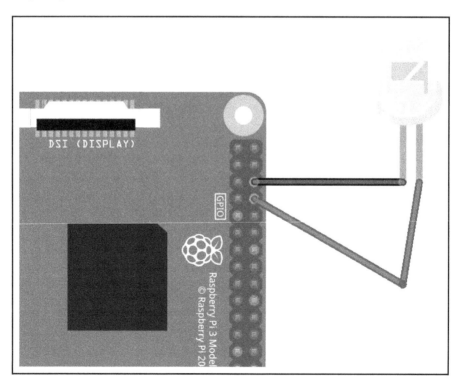

After connecting the led to the RPi, it's time to write the program. The program for blinking the led using the softPWM library is as follows. This program is called `Soft_PWM_Blink.cpp`, and you can download this program from the `Chapter02` folder of this book's GitHub repository:

```cpp
#include <iostream>
#include <wiringPi.h>
#include <softPwm.h>
int main(void)
{
 wiringPiSetup();
 softPwmCreate (15, 0, 100) ;
 for(;;)
 {
 softPwmWrite (15, 25);
 delay(1000);
 softPwmWrite (15, 0);
 delay(1000);
 softPwmWrite (15, 50);
 delay(1000);
 softPwmWrite (15, 0);
 delay(1000);
 softPwmWrite (15, 100);
 delay(1000);
 softPwmWrite (15, 0);
 delay(1000);
 }
 return 0;
 }
```

The explanation of the preceding program is as follows:

- In this program, we first import the `wiringPi` and `iostream` libraries, along with the `softPwm` library.
- Next, in the `main` function, using the `softPwmCreate` function, we are setting wiringPi pin 15 as the PWM pin. The initial duty cycle value is set to 0 and the max duty cycle value is set to `100`.
- After that, inside the `for` loop, we have six `softPwmWrite` functions, and by using these functions, we turn on the LED at different brightness levels.
- With the `softPwmWrite(15,25)` function code, the LED will be HIGH with 25% brightness. Since the delay is set to 1,000, the LED will be HIGH for 1 second.
- After this, since the duty cycle value is set to 0, the led will be LOW for 1 second in the `softPwmWrite(15 , 0)` function code.

- Next, with the `softPwmWrite(15,50)` command, the LED will be HIGH with 50% brightness for 1 second. After this, we are again turning the LED LOW for 1 second.
- Finally, with the `softPwmWrite(15 , 100)` function code, the LED will be HIGH with 100% brightness for 1 second. Next, we again turn the led OFF for 1 second.
- After writing the code, you can click on the compile button to compile the code and after that, hit the build button to test the code.

This is how we control the brightness of an LED using the softPWM library.

Summary

Congratulations—you have successfully written your first C++ program and run it on your Raspberry Pi! In this chapter, we first installed the `wiringPi` library and understood the wiringPi pin connections for the Raspberry Pi. Next, we wrote a simple C++ program to blink an led. After that, we understood the working of the LDR sensor module and turned the LED on/off depending on the input from the LDR sensor module. After this, we understood PWM and used the softPWM library to write a program to vary the brightness of the led.

In the next chapter, we will look at the different parts that are required to create a car robot. Next, we will understand the workings of DC motors and motor drivers, and learn how to create a car robot. After this, we will write a C++ program to move the robot in different directions.

Questions

1. How many ground pins are there on the Raspberry Pi?

2. In a dark environment, is the resistance of the LDR sensor HIGH or LOW?

3. What command is used for reading values from a sensor?

4. What is the for loop command to make the LED blink six times?

5. What will be the output voltage at a duty cycle at 20%, assuming the maximum voltage is 5V?

Section 2: Raspberry Pi Robotics

2

In this section, you will first develop a car robot. After that, you will gain an understanding of the way in which the L298N motor driver works and how to move the robot in different directions. You will also connect an ultrasonic sensor and LCD module to the robot and create an obstacle-avoiding robot at the end.

The following chapters are included in this section:

- Chapter 3, *Programming the Robot*
- Chapter 4, *Building an Obstacle-Avoiding Robot*
- Chapter 5, *Controlling a Robot Using a Laptop*

3
Programming the Robot

After writing a couple of C++ programs and testing their output on the Raspberry Pi, it's now time to create our very own car robot and make it move forward, backward, left, and right.

In this chapter, we will cover the following topics:

- Choosing a good robot chassis
- Constructing and connecting the robot
- Working with H-bridge
- Moving the robot

Technical requirements

The main hardware requirements for this chapter are the following:

- Robot chassis (the parts included in the robot chassis are explained in the *Constructing and connecting the robot* section)
- Two DC motors
- L298N motor driver
- Female-to-female connecting wires

The code files for this chapter can be downloaded from `https://github.com/PacktPublishing/Hands-On-Robotics-Programming-with-Cpp/tree/master/Chapter03`.

Choosing a robot chassis

Choosing a good robot chassis is one of the most important activities to do before we start constructing the robot. The chassis for the robot is like a skeleton for a human. Our skeletons are made up of bones that provide proper support to our organs. In the same way, a good chassis will provide proper support to the electronics components and hold them together.

You can either purchase a robot chassis from e-commerce websites such as Amazon and eBay, or you can purchase one directly from a vendor who deals with robotics equipment. A quick Amazon search for `robot chassis` will provide you with a list of different variants of robot chassis. Choosing from all of these options can be a daunting task if you haven't constructed a robot previously. While choosing a robot chassis, keep the following pointers in mind:

- Make sure that the robot chassis consists of two plates (an **Upper Plate** and a **Lower Plate**) so that you can place the electronics components in between the two plates as well as on the **Upper Plate**, as shown in the following photo:

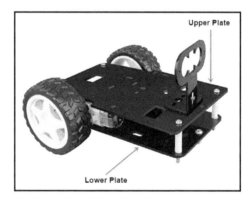

- Choose a robot chassis that only supports two DC motors, like the one shown in the preceding photo. Robot chassis with support for four DC motors are also available, but you would need an extra motor driver to drive a four-wheel robot.
- Finally, choose a robot chassis that has DC motors (two units), wheels (two units), and a castor wheel included as part of the complete kit, so that you do not have to purchase these components separately.

The robot chassis shown in the preceding photo is the one that I'll be using for creating my car robot, as it consists of two plates and includes the necessary components (DC motors, wheels, castor wheels, screws, and spacers) as a part of the complete kit.

Constructing and connecting the robot

Constructing the robot properly is one of the most important steps. A properly constructed robot will move smoothly without any obstructions. Before constructing the robot, let's take a look at the complete list of components that you will need.

The parts required for building the robot include the following:

- A robot chassis, which must include the following components:
    - An upper plate and a lower plate
    - Two BO DC motors (BO is a type of DC motor that generally is yellow in color)
    - **Two** wheels
    - **One** castor wheel
    - Spacers
    - Screws for connecting different parts
- **One** screwdriver
- **One** L298N motor driver
- **Seven or eight** connecting wires
- **One** battery snapper
- **One** 9V battery

Since these robot chassis are created by small-scale companies and there isn't a standard robot chassis that is available internationally, the robot chassis that I'm using for this project will differ from the robot chassis that is available in your country.

 While purchasing a robot chassis online, please check the user reviews for the product.

Constructing the robot

Constructing the robot becomes much easier when components such as upper and lower plates, DC motors, wheels, castor wheels, and spacers are included inside one single robot chassis kit. If you purchase these components separately, there is a chance that some components will not fit properly, which makes the entire assembly of the robot unstable. While the chassis that I'm using may differ from the one that you are using, the construction of most two-wheeled robots is quite similar.

You can check the robot construction in the `Chapter03` folder of the GitHub repository.

Connecting the motor driver to the Raspberry Pi

After constructing the robot, it's time to connect the Raspberry Pi to the motor driver so that we can program the robot and move it in different directions. Before doing this, however, let's take a look at what a motor driver is.

What is a motor driver?

A motor driver is a breakout board that consists of a motor driver **integrated circuit** (**IC**). A motor driver is basically the same as a current amplifier, and its main purpose is to take a low-current signal and convert to a high-current signal in order to run the motors. The L298N motor driver is shown in the following photo:

The main reason why we need a motor driver is that components such as motors cannot be connected directly to the Raspberry Pi as they do not get sufficient current from the Raspberry Pi, as shown in the following diagram:

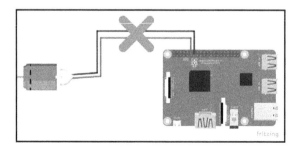

This is why we connect the motor to the motor driver first and supply power to the motors with a battery, as shown in the following diagram:

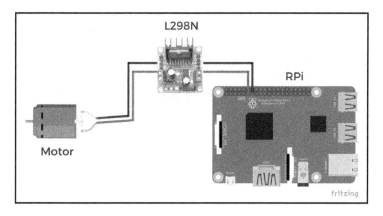

Wiring connections

The L298N motor driver consists of **four** input pins, **four** output sockets (two sockets for each motor), and **two** sockets for power. The Raspberry Pi pins are connected to the input pins of the motor driver. The DC motor wires are connected to the output socket of the motor driver and the battery snapper is connected to the power socket. The four input pins of the L298N motor driver are labeled **IN1**, **IN2**, **IN3**, and **IN4**. The output sockets are labeled **OUT1**, **OUT2**, **OUT3**, and **OUT4**. The following figure shows the wiring connection of Raspberry Pi, motor driver, and motors:

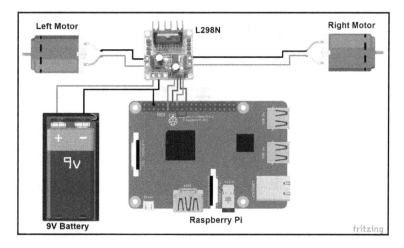

As you can see in the preceding diagram, the wiringPi pin numbers **0**, **2**, **3**, and **4** are connected to the input socket of the motor driver, as follows:

- wiringPi no **0** is connected to **IN1**
- wiringPi no **2** is connected to **IN2**
- wiringPi no **3** is connected to **IN3**
- wiringPi no **4** is connected to **IN4**
- The left motor wires are connected to the **OUT1** and **OUT2** sockets
- The right motor wires are connected to the **OUT3** and **OUT4** sockets
- The red wire of the battery snapper is connected to the **VCC** socket of the motor driver and the black wire is connected to the ground socket
- The ground pin from the Raspberry Pi is connected to the ground socket

Working with H-bridge

The L298N motor driver IC can control two motors at a time. It consists of a dual H-bridge circuit. This means it consists of two circuits that look like the one shown in the following diagram, one for each motor:

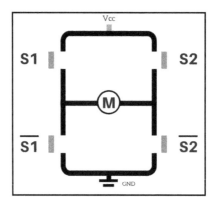

The H-bridge circuit consists of four switches **S1**, **S2**, $\overline{S1}$, and $\overline{S2}$. These switches will open and close based on the input that we provide to the L298N IC.

Now, since we have two motors, there are four possible input combinations that we can provide to the L298N IC, as follows:

- HIGH HIGH (1, 1)
- HIGH LOW (1, 0)

- LOW HIGH (0, 1)
- LOW LOW (0, 0)

We will provide the HIGH (1) and LOW (0) signal to the **S1** and **S2** switches, as follows:

1. First, when $S1 = 1$ and $S2 = 0$, the **S1** switch will be closed and the **S2** switch will remain open. $\overline{S1}$, or $\overline{1}$, will be 0, so the $\overline{S1}$ switch will be open. $\overline{S2}$, or $\overline{0}$, will be 1, so the $\overline{S2}$ switch will be closed. Now, since the **S1** and $\overline{S2}$ switches are closed, the current will flow from the **Vcc** to the **S1**, then to the motor, then to the $\overline{S2}$, and end at the **GND**. The motor will rotate in a clockwise direction, as shown in the following diagram:

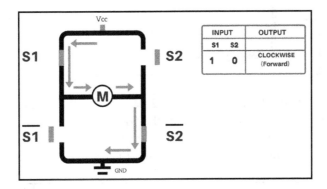

2. When $S1 = 0$ and $S2 = 1$, the **S1** switch will be open, the **S2** switch will be closed, the $\overline{S1}$ will be closed, and the $\overline{S2}$ will be open. Now, since the **S2** and $\overline{S1}$ switches are closed, the current will flow from the **Vcc** to the **S2**, then to the motor, then to the $\overline{S1}$, and end at the **GND**. The motor will rotate in an anticlockwise direction, as shown in the following diagram:

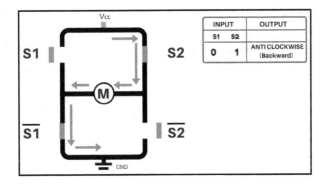

3. When *S1* = 0 and *S2* = 0, the **S1** switch will be open, the **S2** switch will be open, the $\overline{S1}$ switch will be closed, and the $\overline{S2}$ switch will be closed. Now, since both the **S1** and **S2** switches are open, there is no path for the current to flow toward the motors. In this case, the motor will stop, as shown in the following diagram:

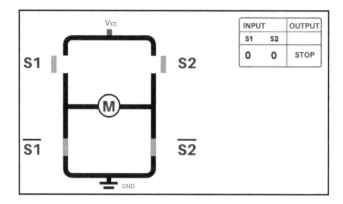

4. When *S1* = 1 and *S2* = 1, the **S1** and **S2** switches will be closed, while the $\overline{S1}$ and $\overline{S2}$ switches will be open. Since both the **S1** and **S2** switches are closed, this will create a short circuit condition and the current will not move through the motor. In this case, the motor will stop, as in the previous case:

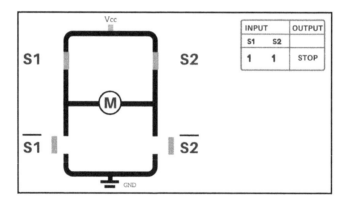

As explained earlier, since the L298N IC consists of two H-bridges, the same process will take place in the other H-bridge when we provide high and low signals. The second H-bridge will control the other motor.

Moving the robot

Now that we have understood the H-bridge circuit, we will write a program called
`Forward.cpp` to move our robot forward. After that, we will write a program to move the
robot backward, left, and right, and then stop. You can download the `Forward.cpp`
program from `Chapter03` of the GitHub repository.

The program for moving the robot forward is as follows:

```
#include <stdio.h>
#include <wiringPi.h>

int main(void)
{
wiringPiSetup();
pinMode(0,OUTPUT);
pinMode(2,OUTPUT);
pinMode(3,OUTPUT);
pinMode(4,OUTPUT);


  for(int i=0; i<1;i++)
  {
digitalWrite(0,HIGH); //PIN O & 2 will move the Left Motor
digitalWrite(2,LOW);
digitalWrite(3,HIGH); //PIN 3 & 4 will move the Right Motor
digitalWrite(4,LOW);
delay(3000);
  }
return 0;
  }
```

Let's see how this program works:

1. First, we set the wiringPi pins (numbers 0, 1, 2, and 3) as output pins.
2. Next, with the following two lines, the left motor moves forward:

    ```
    digitalWrite(0,HIGH);
    digitalWrite(2,LOW);
    ```

3. Then, the next two lines make the right motor move forwards:

```
digitalWrite(3,HIGH);
digitalWrite(4,LOW);
```

4. After that, the `delay` command means the motors will move forwards for three seconds. As we are currently inside a `for` loop, the motor will keep rotating continuously.

5. Once you have finished the code, compile the program to check if there are any errors.

6. Next, connect the 9V battery to the battery snapper and upload the program. Before doing this, however, make sure that you lift up the wheels of the robot. This is because, when the robot starts moving, you might get one of the following three outputs:

  - Both the motors move forwards. If you get this output, this means your robot will move forwards once you put it on the ground.
  - One motor moves forwards and the other motor moves backward. If you get this output, interchange the wires of the motor that is moving backward on the motor driver. For example, if the right motor is moving backward, insert the **M3-OUT** wire in the **M4-OUT** socket and the **M4-OUT** wire in the **M3-OUT** socket, as shown in the following diagram:

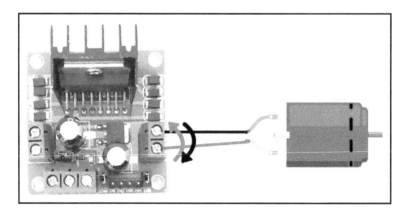

- Both the motors move backward. In this case, your robot will move backward. If you get this output, interchange the wires of both the left and right motor on the motor driver. To do this for the left motor, connect the **M1-OUT** socket wire in the **M2-OUT** socket and the **M2-OUT** socket wire in the **M1-OUT** socket. For the right motor, connect the **M3-OUT** socket wire in the **M4-OUT** socket and the **M4-OUT** socket wire in the **M3-OUT** socket, as shown in the following diagram:

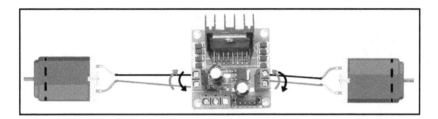

Alternatively you can also interchange the pins on the RPi to move the robot forward; connect pin 0 in place of pin 2 and pin 2 in place of pin 0 for the left motor. Similarly, connect pin 3 in place of pin 4 and pin 4 in place of pin 3 for the right motor.

7. Click on the upload button and check the final output. Since this program is in a `for` loop, the motor will keep running continuously. After testing the output, disconnect the battery from the battery snapper so that the power to the motors through the motor driver is turned off to stop the motors from moving.

Moving the robot backward

To move the robot backward, we simply need to interchange the HIGH signal with the LOW signal and vice versa. The complete program to move the robot in this way is written inside the `RobotMovement.cpp` file, which can be downloaded from `Chapter03` of the GitHub repository:

```
digitalWrite(0,LOW);        //PIN 0 & 2 will move the Left Motor
digitalWrite(2,HIGH);
digitalWrite(3,LOW);        //PIN 3 & 4 will move the Right Motor
digitalWrite(4,HIGH);
delay(3000);
```

The first two lines will make the left motor move backward, while the following two lines will make the right motor move backward. The final line indicates that the robot will move for three seconds.

Stopping the robot

To stop the robot from moving, you can provide a `HIGH` signal or a `LOW` signal to the pins. In the code to make the robot move backward, add the following command to stop the motors for three seconds:

```
digitalWrite(0,HIGH);          //PIN 0 & 2 will STOP the Left Motor
digitalWrite(2,HIGH);
digitalWrite(3,HIGH);          //PIN 3 & 4 will STOP the Right Motor
digitalWrite(4,HIGH);
delay(3000);
```

Different types of turns

There are two types of turn that a robot can carry out:

- Axial turn
- Radial turn

 The code for taking axial and radial turns is added in the `RobotMovement.cpp` program.

Axial turns

In axial turns, one wheel of the robot moves backward and the other wheel of the robot moves forwards. The robot can turn on the spot without moving from its original position. Axial turns are generally carried out if there are space constraints while turning, such as if a robot is moving through a maze. A robot can either carry out an axial left turn or an axial right turn.

Axial left turn

In an axial left turn, the left motor of the robot moves backward and the right motor moves forward, so the robot takes a left turn, as shown in the following diagram:

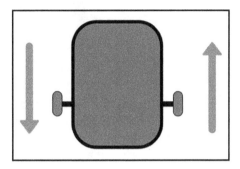

If you have understood how the H-bridge works, you might be able to guess the code for taking an axial turn. If not, the code is as follows:

```
digitalWrite(0,LOW);
digitalWrite(2,HIGH);
digitalWrite(3,HIGH);
digitalWrite(4,LOW);
delay(500);
```

 You will need to play around with the delay value a little bit to make sure that the robot is turning properly in the left direction. If the delay value is high, the robot will turn more than 90°, whereas if it is low, the robot will turn less than 90°.

Axial right turn

In an axial left turn, the left motor of the robot moves forwards and the right motor moves backward, thereby taking a right turn, as shown in the following diagram:

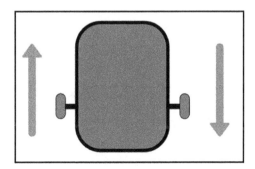

The code for an axial right turn is as follows:

```
digitalWrite(0,HIGH);
digitalWrite(2,LOW);
digitalWrite(3,LOW);
digitalWrite(4,HIGH);
delay(500);
```

Radial turn

In a radial turn, one motor of the robot stops and the other motor moves forwards. The wheel that is stopped acts as the center of the circle and the moving wheel acts as the circumference. The distance between the motors represents a radius, which is why this turn is called a radial turn. A robot can either carry out a radial left turn or a radial right turn.

Radial left turn

In a radial left turn, the left motor stops and the right motor moves forwards, so the robot takes a left turn, as shown in the following diagram:

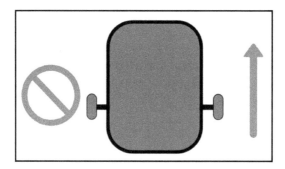

The code for taking a radial left turn is as follows:

```
digitalWrite(0,HIGH);
digitalWrite(2,HIGH);
digitalWrite(3,HIGH);
digitalWrite(4,LOW);
delay(1000);
```

Radial right turn

In a radial right turn, the left motor moves forwards and the right motor stops, so the robot takes a right turn, as shown in the following diagram:

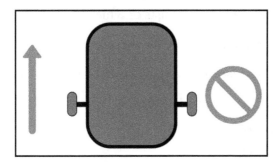

The code for taking a radial right turn is as follows:

```
digitalWrite(0,HIGH);
digitalWrite(2,LOW);
digitalWrite(3,HIGH);
digitalWrite(4,HIGH);
delay(1000);
```

Summary

In this chapter, we have looked at certain criteria for selecting a robot chassis. After that, we constructed our car, connected the motor driver to the Raspberry Pi, and understood the workings of an H-bridge circuit. Finally, we wrote programs to make the robot move forward, backward, left, and right.

In the next chapter, after understanding the basic fundamentals of moving a robot in this chapter, we will first write a program to measure distance using an ultrasonic sensor. Next, we will use these distance values for avoiding obstacles, that is, if the robot comes very close to a wall then the ultrasonic sensor will sense it and it will command the robot to take a turn, thereby avoiding the obstacle.

Questions

1. Which motor driver are we using for controlling the robot?

2. An L298N motor driver IC consists of which bridge?

3. What is the C program to move the robot in the forward direction?

4. *S1 = 0* (LOW) and *S2 = 1* (HIGH), will move the robot in which direction?

5. What is the code for a radial left turn?

6. What is the code for an axial right turn?

Building an Obstacle-Avoiding Robot

4

Now that we can move the robot in multiple directions for specified periods of time, let's think about how to read values from an ultrasonic sensor in order to create a robot that can avoid obstacles. We will also use an LCD display and use it to print distance values.

In this chapter, we will cover the following topics:

- Using an ultrasonic sensor
- Using an LCD
- Creating an obstacle-avoiding robot

Technical requirements

The main hardware requirements for this chapter are the following:

- An HC-SR04 ultrasonic sensor
- A 16x2 LCD or a 16x2 LCD with an I2C LCD module
- A breadboard
- One 1 KΩ resistor
- One 2 KΩ resistor
- 12-13 connecting wires

The code files for this chapter can be downloaded from `https://github.com/PacktPublishing/Hands-On-Robotics-Programming-with-Cpp/tree/master/Chapter04`.

Using an ultrasonic sensor

An ultrasonic sensor is used to measure the distance between an obstacle or an object. An ultrasonic sensor consists of a transmitting transducer and a receiving transducer. The transmitting transducer (the trigger) emits **ultrasonic pulses** (also referred to as **ultrasonic sound**), which collide with nearby obstacles and are received by the receiving transducer (the echo). The sensor determines the distance between a target by measuring the time difference between the sending and the receiving of ultrasonic waves. The following diagram illustrates this process:

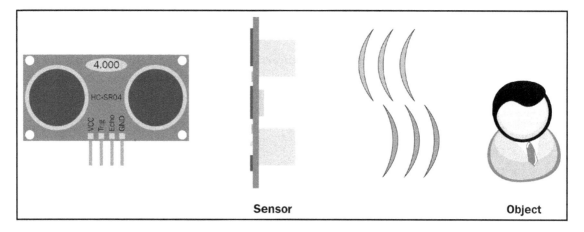

The ultrasonic sensor we will use for this project is called an **HC-SR04 ultrasonic sensor**, which is one of the most widely-used ultrasonic sensors. It can measure distances in the range of 0–180 cm, with a resolution of about 0.3 cm. It has a frequency of around 40 KHz. The HC-SR04 sensor consists of the following four pins:

- VCC pins
- A ground pin
- A trigger pin
- An echo pin

A trigger pin is connected to the transmitting transducer, which transmits pulses, and an echo pin is connected to the receiving transducer, which receives the pulses, as shown in the photo of an HC-SR04 ultrasonic sensor:

How an ultrasonic sensor measures distances

Now that we've understood the basic workings of an ultrasonic sensor, let's think about exactly how the ultrasonic sensor measures distance:

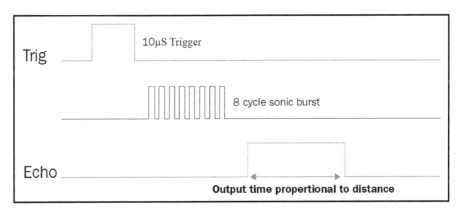

In order to measure distance, the ultrasonic sensor generates an ultrasonic pulse. To generate this ultrasonic pulse, the trigger pin is set in a **high** state for **10 microseconds**. This produces an *eight-cycle sonic burst* that travels at the *speed of sound*, which is received by the echo pin after colliding with an object. When this *eight-cycle sonic burst* is received, the echo will become high and it will remain high for a time duration that is proportional to the time taken for the ultrasonic pulse to reach the echo pin. If it took 20 microseconds for the ultrasonic pulse to reach the echo pin, the echo pin would remain high for 20 microseconds.

The arithmetic equation for determining the time taken

Let's firstly look at the arithmetic equation for calculating distance, which is shown in the following diagram:

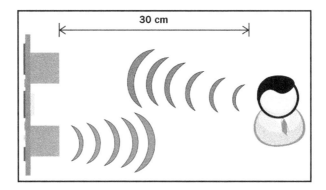

As indicated in the preceding diagram, let's imagine that the distance between the sensor and the object is 30 cm. The ultrasonic sensor travels at a speed of sound, which is 340 m/s, or 0.034 cm/µs.

To calculate the time taken, we will use the following equation:

$$speed = \frac{distance}{time}$$

If we move time to the left-hand side, and speed to the right-hand side, we get the following equation:

$$time = \frac{distance}{speed}$$

If we input the preceding numbers, we get the following:

$$time = \frac{30}{0.034}$$

The result of this equation is that the time taken is 882.35 µs.

Even though the time value is 882.35 µs, the time duration for which the echo pin remains high will actually be double 882.35 µs, which is 1764.70 µs. This is because the ultrasonic sound first travels toward the object, and is received by the echo after bouncing back from the object. It travels the same distance twice: first from the sensor to the object, and then from the object to the sensor. If the time value is doubled, the distance value will also be doubled. We can modify the preceding equation to find the distance as follows:

$$time = \frac{2 X distance}{speed}$$

$$distance = \frac{speed X time}{2}$$

$$distance = \frac{0.034 X time}{2}$$

Make note of this equation, as we will use it later on to find the distance, once we get the time duration value.

Wiring the ultrasonic sensor to the Raspberry Pi

The HC-SRO4 sensor consists of four pins: **VCC**, **GND**, **trigger** (**Trig**), and **echo**, so the wiring connections of the RPi and the ultrasonic sensor should be as follows:

- Connect the **VCC** pin of the sensor to pin number 4.
- Connect the **GND** pin of the sensor to pin number 9.
- Connect the **Trig** pin of the sensor to wiringPi pin number 12.

- The **echo** pin of the sensor is connected to wiringPi pin number 13 via a voltage divider. The two resistors used in the voltage divider circuit have a resistance value of 1 KΩ (**R1**) and 2KΩ (**R2**), respectively. The voltage divider circuit is used to reduce the incoming 5V signal from the echo pin (to the RPi) to 3.3V. The wiring connection of RPi and HC-SR04 is shown in the following figure:

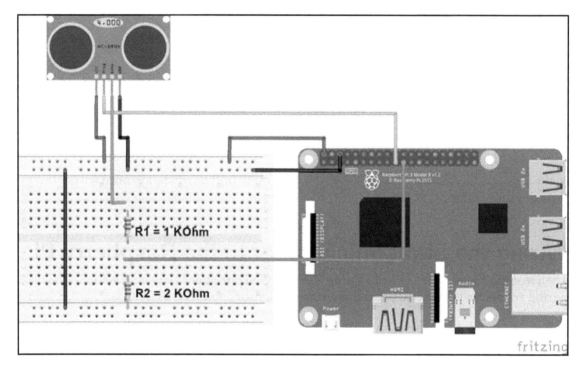

The formula used to convert the incoming voltage to 3.3V is as follows:

$$Vout = VinX\frac{R2}{R2 + R1}$$

Vin is the incoming voltage from the echo pin, **R1** is the first resistor, and **R2** is the second resistor. **Vin** is 5V, **R1** is 1 KΩ, and **R2** is 2 KΩ:

$$Vout = 5X\frac{2K}{2K + 1K}$$

$$Vout = 3.3V$$

The HC-SR04 sensor program

After wiring up the HC-SR04 sensor to the Raspberry Pi, let's write a program for measuring the distance between an object from an ultrasonic sensor. The distance measurement program is called `DistanceMeasurement.cpp` and you can download it from the `Chapter04` folder of the GitHub repository.

The code for measuring the distance is as follows:

```cpp
#include <stdio.h>
#include <iostream>
#include <wiringPi.h>

using namespace std;

#define trigger 12
#define echo 13

long startTime;
long stopTime;

int main()
{

  wiringPiSetup();

  pinMode(trigger,OUTPUT);
  pinMode(echo, INPUT);

for(;;){
  digitalWrite(trigger,LOW);
  delay(500);

  digitalWrite(trigger,HIGH);
  delayMicroseconds(10);

  digitalWrite(trigger,LOW);

  while(digitalRead(echo) == LOW);
  startTime = micros();

  while(digitalRead(echo) == HIGH);
  stopTime = micros();

long totalTime= stopTime - startTime;
  float distance = (totalTime * 0.034)/2;
```

```
cout << "Distance is: " << distance << " cm"<<endl;
delay(2000);
}
return 0;
}
```

In the preceding code, we declared the `wiringPi`, `stdio`, and `iostream` libraries. After that, we declared the `std` namespace:

1. After this, with the lines `#define trigger 12` and `#define echo 13`, we declare wiringPi pin number 12 as the trigger pin and wiringPi pin number 13 as the echo pin.

2. Then, we declare two variables called `startTime` and `stopTime`, which are of the datatype `Long`. The `startTime` variable will record the time when the ultrasonic pulse is sent by the trigger pin and the `stopTime` variable will record the time when the ultrasonic pulse is received by the echo pin.

3. Inside the main function, the trigger pin is set up as `OUTPUT`, as it will generate the ultrasonic pulse. The echo pin is set up as `INPUT`, as it will receive the ultrasonic pulse.

4. Inside a `for` loop, we set the trigger pin to `LOW` for 500 milliseconds, or 0.5 seconds.

5. To generate the ultrasonic pulse, the trigger pin is set to `HIGH` (`digitalWrite(trigger,HIGH)`) for 10 microseconds (`delayMicroseconds(10)`). After generating the pulse for 10 µs, we set the trigger pin to `LOW` again.

6. Next, we have two `while` loops, inside of which, there are two `micros()` functions. The `micros()` will return the current time value in milliseconds. The first while loop (`digitalRead(echo) == LOW`) will record the time at the beginning of the pulse, and the time duration in which the echo pin is `LOW` will be stored in the `startTime` variable.

7. When the pulse is received by the echo pin, the second `while` loop (`digitalRead(echo) == HIGH`) will execute. The `micros()` function inside this while loop will return the time value for the time taken for the ultrasonic pulse to reach the echo pin. This time value will be stored in the `stopTime` variable.

8. Next, to find the total time, we subtract the `startTime` from the `stopTime` and store this time value in the `totalTime` variable.

9. After finding out the `totalTime`, we use the following formula to calculate the distance:

$$float\ distance = (totalTime \times 0.034)/2$$

10. To display the distance value, we will use the `cout` statement. The `delay(2000);` command is called so that the distance value is printed after every two seconds.

After completing the code, you can compile and build it to check the final output. You can place an object in front of the sensor and the object's distance from the sensor will be displayed inside the console.

On my robot's chassis, there is an additional part on which I have fixed the ultrasonic sensor.

Using an LCD

A **liquid crystal display** (**LCD**) is an electronic display unit that is generally used in computers, TVs, smartphones, and cameras. A 16x2 LCD is a basic LCD module that is generally used in electronics or DIY projects. As the name suggests, a 16x2 LCD consists of 16 columns and 2 rows. This means that it has two lines, on each of which we can display a maximum of 16 characters. A 16x2 LCD consists of 16 pins labeled from **VSS** to **K**, as shown in the following photo:

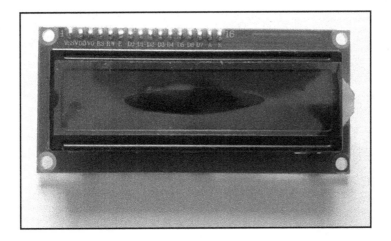

Each pin on the LCD can be described as follows:

Pin number	Name	How it works
1	VSS (GND)	Ground pin.
2	VCC	The VCC pin requires 5V of power in order to turn on the LCD module.
3	Vo	Using this pin, we can adjust the contrast of LCD. We can connect it to GND for maximum contrast. If you want to vary the contrast, connect it to the data pin of a potentiometer.
4	RS (Register Select)	The LCD consist of two registers: the command register and the data register. The RS pin is used to switch between the command and the data register. It is set to HIGH (1) for the command register and LOW (0) for the data register.
5	R/W (Read Write)	Set this pin to LOW to write to the register, or set it to HIGH to read from the register.
6	E (Enable)	This pin enables the clock of the LCD, so that the LCD can execute instructions.
7	D0	Even though the LCD has eight data pins, we can either use it in eight-bit mode or four-bit mode. In eight-bit mode, all the eight data pins (D0-D7) are connected to the RPi pins. In four-bit mode, only four pins (D4-D7) are connected to the RPi. We will use the LCD in four-bit mode in this case, so that fewer wiringPi pins are occupied.
8	D1	
9	D2	
10	D3	
11	D4	
12	D5	
13	D6	
14	D7	
15	A (Anode)	+5V pin for LCD backlight.
16	K (Cathode)	GND pin for LCD backlight.

Since a 16x2 LCD has a total of 16 pins, connecting all the pins correctly to the Raspberry Pi can sometimes be an issue. If you make a mistake and a pin that needs to be connected to D0 gets connected to D1, for example, you might get incorrect output.

To avoid this potential confusion, you can choose to purchase an **I2C LCD adapter module** for a 16x2 LCD. This module takes the 16 pins of LCD as an input, and provides only 4 pins as an output (VCC, GND, SDA, SCL). This means that you only need to connect 4 pins to the Raspberry Pi, instead of 16 pins.

The 16x2 LCDs with I2C LCD adapters soldered to them are also available, which can save you some time. The 16x2 LCD that I'm using for this project already has an I2C LCD adapter soldered to it, as shown in the following picture:

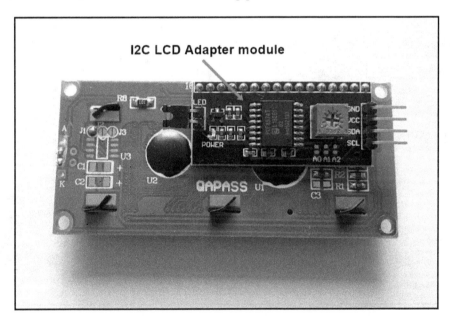

In the following sections, we'll understand the wiring connections and how to program both a normal LCD and an LCD with an I2C LCD adapter.

I will refer to the **16x2 LCD with the I2C LCD adapter** as I2C LCD to avoid complication.

Wiring the 16x2 LCD to the Raspberry Pi

To connect the 16x2 LCD to the Raspberry Pi, you will need a mini breadboard, as there are a couple of pins that need to be connected to the VCC and the GND. The wiring connections of RPi and the 16x2 LCD are as follows:

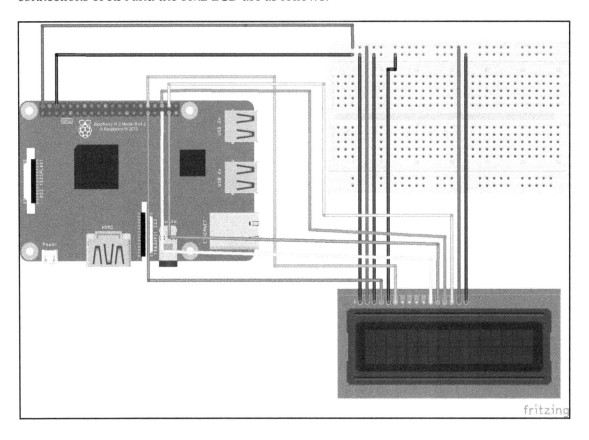

First, connect pin number 2 or pin number 4 from the Raspberry Pi to one horizontal pin of the breadboard, so that we can use that row as a VCC row. Similarly, connect one ground pin from the Raspberry Pi to a horizontal pin of the breadboard, so that we can use that row as the ground row. Next, follow these instructions:

1. Connect the VSS (GND) pin to the ground row of breadboard
2. Connect the VCC pin to the VCC row of the breadboard
3. Connect the V0 pin to the ground row of breadboard
4. Connect the **register select** (**RS**) pin to wiringPi pin number 22 of the RPi
5. Connect the R/W pin to the ground row of the breadboard, as we will write off the LCD's register
6. Connect the enable pin to wiringPi pin number 26 of the RPi
7. We will use the LCD in four-bit mode, so pins D0 to D3 will remain unconnected
8. Pin D4 should be connected to wiringPi pin number 24 of the RPi
9. Pin D5 should be connected to wiringPi pin number 25 of the RPi
10. Pin D6 should be connected to wiringPi pin number 27 of the RPi
11. Pin D7 should be connected to wiringPi pin number 28 of the RPi
12. Connect the anode pin to the VCC row of the breadboard
13. Connect the cathode pin to the ground row of the breadboard

For testing the LCD program, add `-lwiringPiDev` command inside **Compile** and **Build** option by opening **Build | Set Build Commands** as shown in the following screenshot:

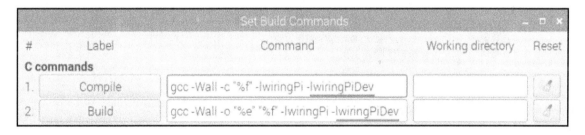

After connecting the 16X2 LCD to the RPi, let's program the LCD.

Programming the LCD

We will write two programs using the normal 16x2 LCD. In the first program, we will print a value on the 16x2 LCD. In the second program, we will print the ultrasonic sensor value on the LCD screen. The first program is called LCDdisplay.cpp and you can download it from the GitHub repository of Chapter04.

The LCD program

After connecting the LCD to the Raspberry Pi, let's examine the program for printing the value on the LCD, as follows:

```
#include <wiringPi.h>
#include <lcd.h>

#define RS 22 //Register Select
#define E 26 //Enable

#define D4 24 //Data pin 4
#define D5 25 //Data pin 5
#define D6 27 //Data pin 6
#define D7 28 //Data pin 7

int main()
{

int fd;
wiringPiSetup();
fd= lcdInit (2, 16, 4, RS, E, D4, D5, D6, D7, 0, 0, 0, 0);
lcdPuts(fd, "LCD OUTPUT");

}
```

The following is the details of the preceding program:

1. First, we call the LCD.h library. The LCD.h library consists of all the important functions that we can use to print, position, and move text, as well as clear the LCD screen.
2. Next, we define pin numbers RS, E, D4, D5, D6, and D7.

3. After this, inside the `lcdInit` function, the first number, which is 2, represents the number of rows in the LCD, while the number `16` represents the number of columns. The number `4` means that we are using the LCD in four-bit mode. Next, we have the RS and E pins, and finally, we have the four data pins. Since we haven't connected the D0, D1, D2, and D3 data pins to the RPi, we have four zeros at the end.

4. The `lcdPuts` is used to print data on the LCD. It takes two parameters as input: the `fd` variable and the text value that needs to be displayed.

5. After completing this code, you can compile and build the code to test the final output.

6. In the output, you will notice that the text output will start from column one, instead of column zero.

7. To position the text on the extreme left side, or column 0, row 0, we need to use the `lcdPosition()` function. The `lcdPosition(fd, column position, row position)` function consists of three parameters, and it should be written before the `lcdPuts` function as follows:

```
fd= lcdInit (2, 16, 4, RS, E, D4, D5, D6, D7, 0, 0, 0, 0);
lcdPosition(fd, 0, 0);
lcdPuts(fd, "LCD OUTPUT");
```

 If the text is not positioned at column 0 and row 0, restart your RPi and test the code once again.

The LCD and the ultrasonic sensor program

After printing a simple text value on the LCD, let's take a look at how to view the ultrasonic distance value on the LCD screen. The wiring connection of the HC-SR04 ultrasonic sensor, the 16x2 LCD, and the RPi are as follows:

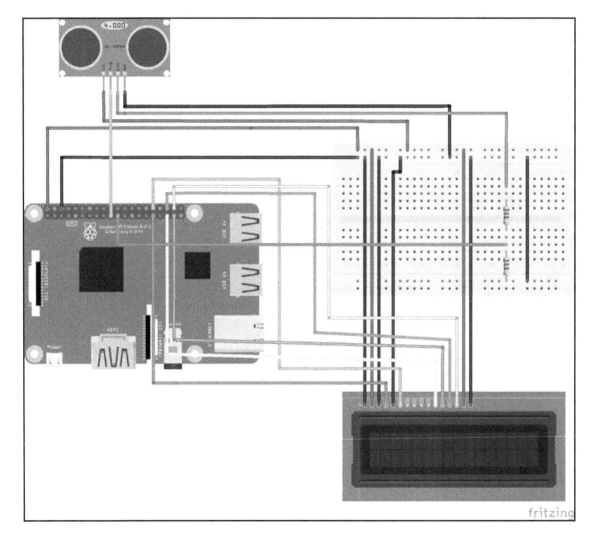

The LCD connections to the RPi remain the same. The ultrasonic trigger pin is connected to wiringPi pin number 12 and the echo pin is connected to wiringPi pin number 13. Let's now take a look at the program. This program is called `LCDdm.cpp` (**dm** is short for **distance measurement**) and you can download it from the GitHub repository of `Chapter04`. The `LCDdm.cpp` program is a combination of the `LCDdisplay.cpp` and `DistanceMeasurement.cpp` programs:

```
int main()
{
...
for(;;)
{
...
cout << "Distance is: " << distance << " cm"<<endl;
lcdPosition(fd, 0, 0);          //position the cursor on column 0, row 0
lcdPuts(fd, "Distance: ");      //this code will print Distance text
lcdPosition(fd, 0, 1);          //position the cursor on column 0, row 1
lcdPrintf(fd, distance);        // print the distance value
lcdPuts(fd, " cm");
delay(2000);
clear();
}
return 0
}
```

In the preceding code, after finding out the distance value, we position the cursor in row zero, column zero, using the `lcdPosition(fd, 0, 0);` command. Next, with the `lcdPuts(fd, "Distance: ")` code, we are displaying the distance text. After this, we position the cursor in column zero and row one. Finally, to print the distance value, we use the `lcdPrintf(fd, distance);` command. Since we have set the delay to two seconds, the distance value will be printed every two seconds. It will then be cleared (`clear()`) and replaced with a new value.

What is the I2C protocol?

The I2C protocol is used in many electronic devices. We use it to connect one master device to multiple slave devices, or multiple master devices to multiple slave devices. The main advantage of the I2C protocol is that the master needs only two pins to communicate with multiple slave devices.

In an I2C bus, all devices are connected in parallel to the same two-wire bus. We can connect a total of 128 devices using 7-bit addressing, and a total of 1,024 devices using 10-bit addressing, as shown in the following diagram:

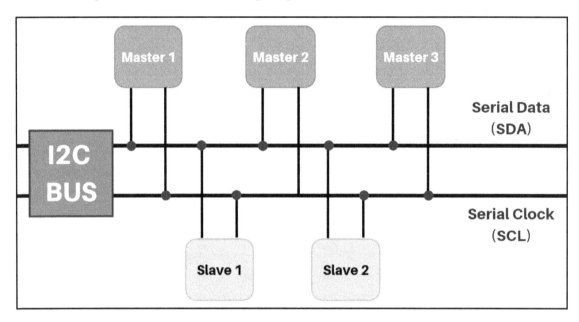

Each device connected using the I2C protocol has a unique ID, which makes it possible to communicate with multiple devices. The two main pins in the I2C protocol are the **Serial Data (SDA)** pin and the **Serial Clock (SCA)** pin:

- **SDA**: The SDA line is used for transferring data.
- **SCL**: The SCL is generated by the master device. It is a clock signal that synchronizes the data transfer between the devices connected in the I2C.

Now that we've understood the basics of the I2C protocol, let's look at how to connect the I2C LCD and the Raspberry Pi.

Wiring the I2C LCD and the Raspberry Pi

On the Raspberry Pi, physical pin number 3 is the SDA pin, while physical pin number 5 is the SCA pin, as shown in the following diagram:

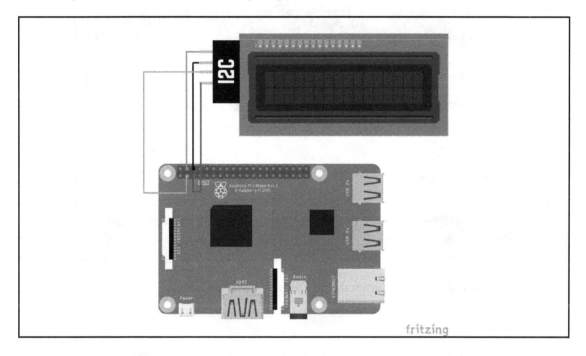

The following is the details of connecting LCD with the RPi:

1. Connect pin number 3 of the Raspberry Pi to the SDA pin of the LCD
2. Connect pin number 5 of the Raspberry Pi to the SCA pin of the LCD
3. Connect the GND pin of the LCD to the GND pin of the RPi
4. Connect the VCC pin of the LCD to pin number 2 or pin number 4 of the Raspberry Pi

Programming the LCD with the I2C LCD module

Before writing the program, we first need to enable the I2C protocol from the Raspberry Pi configuration. To do this, open the command window and type in the following command:

```
sudo raspi-config
```

Inside configurations, open **Interfacing Options**, shown as follows:

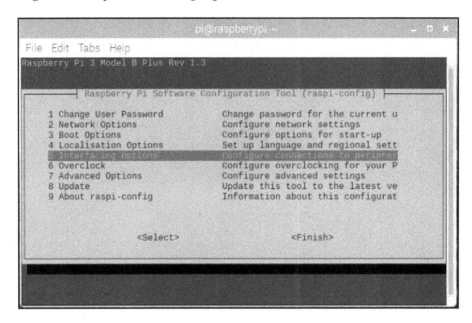

Next, open the **I2C** option, as shown in the following screenshot:

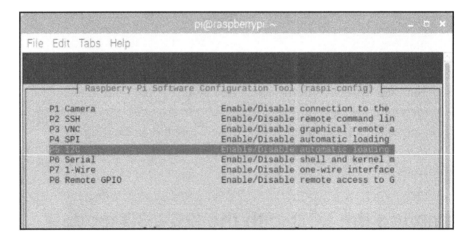

Select the **Yes** option and press *Enter* to enable I2C, as follows:

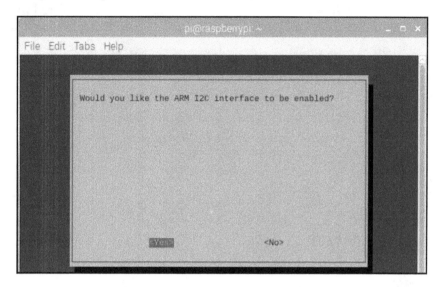

After enabling I2C, select the **Ok** option and exit the configuration, shown as follows:

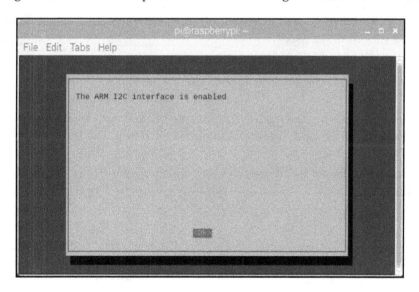

After enabling the I2C protocol inside your Raspberry Pi, let's write the program to print values to the LCD. The program is called I2CLCD.cpp and you can download it from the GitHub repository of Chapter04.

Since this LCD has an I2C module attached to it, the `LCD.h` library that we used in the previous LCD program will not work in this program. Instead, I have created five main functions that will initialize the LCD, print messages, and clear the LCD screen, as follows:

- `init_lcd()`: This function will initialize (set up) the LCD
- `printmessage()`: This function is used to print a string on the LCD
- `printInt()`: This function is used to display integer values
- `printfloat()`: This function is used to display float values
- `clear()`: This function will clear the LCD screen

```
#include <wiringPiI2C.h>
#include <wiringPi.h>
#include <stdlib.h>
#include <stdio.h>

#define I2C_DEVICE_ADDRESS 0x27
#define firstrow 0x80 // 1st line
#define secondrow 0xC0 // 2nd line
int lcdaddr;
```

1. We start the program by declaring the `wiringPiI2C.h` library. Next, we have the `wiringPi` library and two other libraries, which are standard C libraries.
2. After this, with the `#define I2C_DEVICE_ADDRESS 0x27` command, we define the I2C device address, which is `0x27`.
3. The `0x80` command represents the first row: row zero, column zero. With the `#define firstrow 0x80` command, we initialize the first line of the LCD.
4. Similarly, `0xC0` represents the second row of LCD: row one, column zero. With the `#define secondrow 0xC0` command, we initialize the second line of the LCD.
5. Next, inside the `lcdaddr` variable, we will store the I2C LCD's address, as follows:

```
int main() {

wiringPiSetup();

lcdaddr = wiringPiI2CSetup(I2C_DEVICE_ADDRESS);

init_lcd(); // initializing OR setting up the LCD
for(;;) {

moveCursor(firstrow);
printmessage("LCD OUTPUT");
```

```
moveCursor(secondrow);
printmessage("USING I2C");
delay(2000);
clear();

moveCursor(firstrow);
printmessage("Integer: ");
int iNumber = 314;
printInt(iNumber);

moveCursor(secondrow);
printmessage("Float: ");
float fNumber= 3.14;
printFloat(fNumber);
delay(2000);
clear();
}
return 0;
}
```

6. Inside the `main()` function, we store the device address inside the `lcdaddr` variable.

7. After this, we initialize, or set up the LCD with the `init_lcd();` command.

8. Next, in the `for` loop, we move the cursor to the first row with the `moveCursor(firstrow);` command.

9. Now, since the cursor is in the first row, the `LCD OUTPUT` text inside the `printmessage("LCD OUTPUT"` code will be printed on the first row.

10. The cursor is then moved to the second row with the `moveCursor(secondrow)` command. The `USING I2C` text is printed on that row.

11. The text on the first and second rows will be visible for two seconds, after which time the LCD screen will be cleared with the `clear()` command.

12. After this, with the next four lines, an integer, `314`, will be printed on the first row. The `printInt(iNumber)` function is used to display the integer value.

13. Similarly, the `printFloat(iFloat)` function is used to display the float value. In the next four lines, `float 3.14` will be printed on the second row.

14. After this, we again clear the LCD.

This is how we can display string, numeric, and float values inside our I2C LCD.

The I2C LCD and the ultrasonic sensor program

To read the ultrasonic sensor value inside an I2C LCD, connect the ultrasonic sensor and I2C LCD to the RPi. From the GitHub repository of `Chapter04`, you can download the complete program called `I2CLCDdm.cpp` program. The wiring connections of I2C LCD, ultrasonic sensor, and the RPi is shown in the following figure:

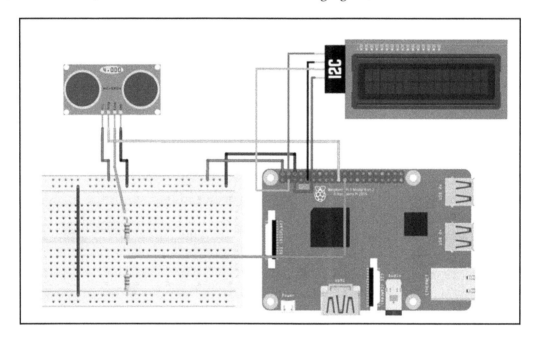

This `I2CLCDdm.cpp` program is basically a combination of the `DistanceMeasurement.cpp` and the `I2CLCD.cpp` programs. In this program, after declaring all the necessary libraries and variables related to the ultrasonic sensor and the I2C LCD below the `cout << "Distance: "<<distance << "cm" << endl` line, we need to add the following code:

```
moveCursor(firstrow);
printmessage("DISTANCE");
moveCursor(secondrow);
printFloat(distance);
printmessage(" cm");
delay(2000);
clear();
```

On the first row, the text DISTANCE will be printed using the printmessage("DISTANCE") command. After that, on the second row, the distance value will be printed using the printFloat(distance) command, since the code is still on the second line. With the printmessage(" cm") command, the cm text will be printed beside the distance value.

The distance value inside the console and the I2C LCD will be visible for two seconds. Next, with the clear() function, the old distance value will be cleared and replaced with a new value. In the console, however, the new value will be displayed on the next line.

Building an obstacle-avoiding robot

In this case, our robot will move freely in a given space, but as soon as it comes near to an object or an obstacle, it will turn or move backward, thus avoiding the obstacle. In this kind of project, we generally use an ultrasonic sensor. As the robot moves, the ultrasonic sensor keeps measuring the distance it is away from objects. When the sensor detects that the distance value is very low, and the robot may collide with the nearby object, it will command the robot to change direction, thus avoiding the obstacle.

To create an obstacle-avoiding robot, you first need to mount the ultrasonic sensor on the robot. Inside my robotic kit, there is already an attachment that allows me to mount the ultrasonic sensor on the robot. This attachment looks as follows:

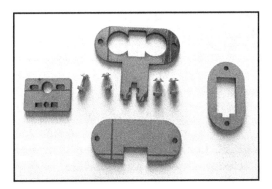

After attaching the ultrasonic sensor on the robot, the final assembly looks as follows:

Wiring connections

The ultrasonic sensor's trigger pin is connected to wiringPi pin number 12, while the echo pin is connected to wiringPi pin number 13 via the voltage divider circuit. The VCC pin of the ultrasonic sensor is connected to physical pin 2 (5V) of RPi, and the ground pin of the ultrasonic sensor is connected to physical pin 6 of RPi. The remaining connections are as follows:

- **WiringPi pin 0** is connected to the **IN1 pin** of the L298N motor driver.
- **WiringPi pin 2** is connected to the **IN2 pin** of the L298N motor driver.
- **WiringPi pin 3** is connected to the **IN3 pin** of the L298N motor driver.
- **WiringPi pin 4** is connected to the **IN4 pin** of the L298N motor driver.
- The **motor driver's ground pin** is connected to the **physical pin 3** of RPi.
- I'm using an I2C LCD, so the **SDA pin** of the I2C LCD is connected to the **physical pin 3 of the RPi**, and the **SCL pin** is connected to the **physical pin 5**. The **ground pin of I2C LCD** is connected to **physical pin 9**, and the **VCC pin of I2C LCD** is connected to **physical pin 4** of RPi.

 Connecting the LCD display to the robot is totally up to you. If you have sufficient space on the robot where the LCD can be placed, go ahead and add it. If not, this is not a necessity.

Programming the obstacle-avoiding robot

In this program, we will first find out the distance of a nearby object using the ultrasonic sensor. Next, we will create an `if` condition that monitors the distance value. If the distance goes below a certain value, we will command the robot to take a turn. Otherwise, the robot will keep moving forward. You can download the complete code called `ObstacleAvoiderRobot.cpp` from `Chapter04` in the GitHub repository:

```cpp
int main()
{
...
for(;;)
{
...
if(distance < 7)
{
digitalWrite(0,LOW);
digitalWrite(2,HIGH);
digitalWrite(3,HIGH);
digitalWrite(4,LOW);
delay(500);
moveCursor(firstrow);
printmessage("Obstacle Status");
moveCursor(secondrow);
printmessage("Obstacle detected");
clear();
}
else
{
digitalWrite(0,HIGH);
digitalWrite(2,LOW);
digitalWrite(3,HIGH);
digitalWrite(4,LOW);
moveCursor(firstrow);
printmessage("Obstacle Status");
moveCursor(secondrow);
printmessage("No Obstacle");
clear();
}
}
return 0;
}
```

In this code, if the **distance** is **greater** than **7 cm**, the robot will keep moving forward. Now, as long as the obstacle is not present, the LCD will display the message `No Obstacle` on the second row. If an obstacle is detected, the robot will first make a radial left turn for 0.5 seconds and the I2C LCD will display the `Obstacle detected` text on the second row. You can increase or decrease the delay value depending on the speed of your motors.

Summary

In this chapter, we have looked at how an ultrasonic sensor works, and we wrote a program to measure the distance values. Next, we programmed the 16x2 LCD, and read the ultrasonic distance value using it. We also looked at the I2C LCD, which takes the 16 LCD pin as an input, and provides four pins as an output, thus simplifying the wiring connections. Finally, we fitted the ultrasonic sensor on our robot to create our obstacle-avoiding robot. This robot moved freely when there was no obstacle near it, and if it approached an obstacle, it would avoid it by taking a turn.

In the next chapter, we are going to create two different types of PC-controlled robot. In the first PC-controlled robot, we will use a library called **ncurses** and use the keyboard as an input. In the second PC-controlled robot, we will create UI buttons using QT, and then use them to move the robot.

Questions

1. What type of pulse does an ultrasonic sensor send?

2. What does LCD stand for?

3. Up to what distance can an HC-SR04 ultrasonic sensor measure?

4. Which row and column would the `lcdPosition(fd, 4,1)` command start printing the text?

5. What are the functions of the anode pin (pin 15) and the cathode pin (pin 16) pin on an LCD?

5
Controlling a Robot Using a Laptop

Controlling a robot with a computer is a fascinating thing. The computer becomes a remote controller, and the robot moves according to the commands provided by the keyboard. In this chapter, we will look at two techniques for controlling a robot wirelessly using your laptop.

We will cover the following topics:

- Installing the `ncurses` library
- Controlling LEDs and a buzzer using `ncurses`
- Controlling a rover (RPi robot) using a laptop keyboard
- Installing and setting up QT5
- Controlling LEDs with GUI buttons
- Controlling a rover using a laptop with QT5

Technical requirements

The main hardware components that you need for this project are as follows:

- Two LEDs
- One buzzer
- A RPi robot

The code files for this chapter can be downloaded from `https://github.com/ PacktPublishing/Hands-On-Robotics-Programming-with-Cpp/tree/master/Chapter05`.

Installing the ncurses library

New curses (ncurses) is a programming library that allows developers to create text-based user interfaces. It is generally used for creating GUI-based applications or software. One of the key features of the ncurses library is that we can use it for taking inputs from keyboard keys, and controlling hardware devices on the output side. We will use the ncurses library to write programs to detect keys to control our robot accordingly. For example, if we press the up arrow, we want our robot to move forward. If we press the left arrow, we want our robot to take a left turn.

To install the ncurses library, we first have to open the command window. To install ncurses, type in the following command and press *Enter*:

```
sudo apt-get install libncurses5-dev libncursesw5-dev
```

Next, you will be asked whether you want to install the library. Type *Y* (for yes) and press *Enter*. It will take around three to five minutes for the ncurses library to download and install inside your RPi.

 Make sure that your RPi is near the Wi-Fi router, so that the library files can download quickly.

ncurses functions

After installing the ncurses library, let's explore some of the important functions that are a part of this library:

- `initscr()`: The `initscr()` function initializes the screen. It sets up the memory, and clears the command window screen.
- `refresh()`: The refresh function refreshes the screen.
- `getch()`: This function will detect the user's touch, and will return the ASCII number for that particular key. The ASCII number is then stored in an integer variable, which is later used for comparison purposes.
- `printw()`: This function is used to print string values inside the command window.

- keypad(): If the keypad function is set to true, we can also take the user's input from the function keys, as well as the arrow keys.
- break: This function is used to exit the program if the program is running in a loop.
- endwin(): The endwin() function frees the memory, and ends ncurses.

The entire ncurses program must be written between the initscr() and endwin() functions:

```
#include <ncurses.h>
...
int main()
{
...
initscr();
...
...
endwin();
return 0;
}
```

Writing a HelloWorld program with ncurses

Let's now write a simple ncurses program for printing Hello World. I have named this program HelloWorld.cpp. The HelloWorld.cpp program can be downloaded from the Chapter05 folder of the GitHub repository:

```
#include <ncurses.h>
#include <stdio.h>

int main()
{
initscr(); //initializes and clear the screen
int keypressed = getch();
if(keypressed == 'h' || keypressed == 'H')
{
printw("Hello World"); //will print Hello World message
}
getch();
refresh();

endwin(); // frees up memory and ends ncurses
return 0;
}
```

The program for compiling and running a C++ program using the `ncurses` library is different from other programs. First, we need to understand the program. After that, we will learn how to compile and run it.

In the preceding code snippet, we first declare the `ncurses` library and the `wiringPi` library. Next, we carry out the following steps:

1. Inside the `main` function, we declare the `initscr()` function to initialize and clear the screen.
2. Next, when the user presses a key, the `getch` function will be called, and the ASCII number of that key will be stored in the `keypressed` variable, which is of the `int` type.
3. After that, using the `for` loop, we check whether the key that is pressed is `'h'` or (`||`) `'H'`. Make sure that you put the letter H in single quotes. When we put the letters in single quotes, we get the ASCII number of that character. `'h'`, for example, returns the ASCII number **104**, while `'H'` returns the ASCII number **72**. Instead of `'h'` or `'H'`, you can also write the ASCII numbers of the *h* and *H* key presses, which are 104 and 72 respectively. This would look as follows: `if(keypressed == 72 || keypressed == 104)`. The numbers should not be inside quotes.
4. Then, if you press the `'h'` or `'H'` key, `Hello World` will be printed inside the command window:

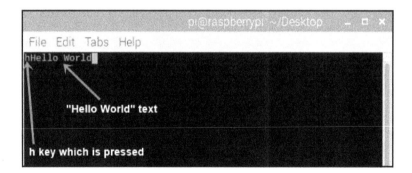

"Hello World" text

h key which is pressed

5. If you want `Hello World` to be printed on the next line, you can simply put `\n` before the `Hello World` text. This would look as follows: `printw("\nHello World")`.
6. After this, when you press a key, the `getch()` function below the `if` condition will be called, and the program will terminate.

Compiling and running the program

For compiling and running the HelloWorld.cpp program, open the Terminal window. Inside the Terminal window, type ls and press *Enter.* You will now see a list of all the folder names present inside your RPi:

The HelloWorld.cpp is stored inside the Cprograms folder. To open the Cprograms folders, type cd (change directory) followed by the folder name, and press *Enter:*

```
cd Cprograms
```

The output of the previous command can be seen as follows:

Next, to view the content of the Cprograms folder, we will type ls again:

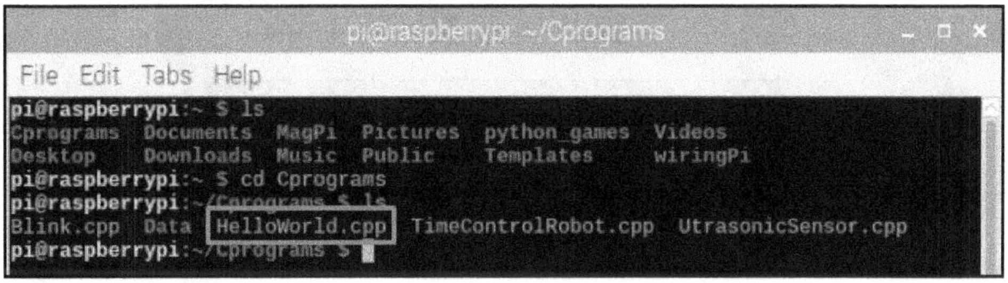

Inside the `Cprograms` folder, there is a `Data` folder and a couple of `.cpp` programs. The program in which we're interested is the `HelloWorld.cpp` program, as we want to compile and build this program. To do this, type the following command and press *Enter:*

```
gcc -o HelloWorld -lncurses HelloWorld.cpp
```

The following screenshot shows that the compilation was done successfully:

For compiling any code that uses the `ncurses` library, the code is as follows:

```
gcc -o Programname -lncurses Programname.cpp
```

After this, type `./HelloWorld` and press *Enter* to run the code:

After you press *Enter*, the entire Terminal window will be cleared:

Next, press the *h* or *H* key, and the `Hello World` text will be printed in the Terminal window. To exit the Terminal window, press any key:

Now that we've created a simple `HelloWorld` program, and tested that the `ncurses` libraries work inside the Terminal window, let's write a program to control the LEDs and the buzzer.

Controlling LEDs and a buzzer using ncurses

After compiling and testing your first `ncurses` program, let's write a program to control hardware such as LEDs and a buzzer by providing input from the keyboard.

Wiring connections

For this particular example, we will need two LEDs and one buzzer. The wiring connections of the LEDs and the buzzer to the RPi are as follows:

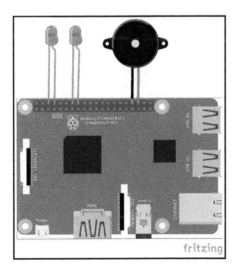

We can see the following from the connection diagram:

- The positive (anode) pin of the first LED is connected to the wiringPi pin number 15, while the negative (cathode) pin is connected to the physical pin number 6 (the ground pin).
- The positive pin of the second LED is connected to the wiringPi pin number 4, while the negative pin is connected to the physical pin number 14 (the ground pin).
- One pin of the buzzer is connected to the wiringPi pin number 27 and the other pin is connected to the physical pin number 34 (the ground pin).

Writing the LEDBuzzer.cpp program

The name of our program is LEDBuzzer.cpp. The LEDBuzzer.cpp program can be downloaded from the Chapter05 folder of the GitHub repository. The LEDBuzzer program is as follows:

```cpp
#include <ncurses.h>
#include <wiringPi.h>
#include <stdio.h>
int main()
{
 wiringPiSetup();

 pinMode(15,OUTPUT); //LED 1 pin
 pinMode(4, OUTPUT); //LED 2 pin
 pinMode(27,OUTPUT); //Buzzer pin

 for(;;){

 initscr();

 int keypressed = getch();

 if(keypressed=='L' || keypressed=='l')
 {
  digitalWrite(15,HIGH);
  delay(1000);
  digitalWrite(15,LOW);
  delay(1000);
 }

 if(keypressed== 69 || keypressed=='e')        // 69 is ASCII number for E.
 {
```

```
digitalWrite(4,HIGH);
delay(1000);
digitalWrite(4,LOW);
delay(1000);
}

if(keypressed=='D' || keypressed=='d')
{
digitalWrite(15,HIGH);
delay(1000);
digitalWrite(15,LOW);
delay(1000);
digitalWrite(4,HIGH);
delay(1000);
digitalWrite(4,LOW);
delay(1000);
}

if(keypressed=='B' || keypressed== 98)          //98 is ASCII number for b
{
digitalWrite(27,HIGH);
delay(1000);
digitalWrite(27,LOW);
delay(1000);
digitalWrite(27,HIGH);
delay(1000);
digitalWrite(27,LOW);
delay(1000);
}

if(keypressed=='x' || keypressed =='X')
{
break;
}

refresh();
}
endwin(); //
return 0;
}
```

After writing the program, let's look at how it works:

1. In the preceding program, we start by declaring the ncurses and wiringPi libraries, along with the stdio C library

2. Next, pin numbers 15, 4, and 7 are declared as output pins

3. Now, when the *L* or *l* keys are pressed, LED 1 will turn HIGH and LOW for one second each

4. Similarly, when the *E* or *e* keys are pressed, LED 2 will turn HIGH and LOW for one second each

5. If *D* or *d* keys are pressed, LED 1 will turn HIGH and LOW for one second each, and then LED 2 will turn HIGH and LOW for one second each

6. If the *b* or *B* keys are pressed, the buzzer will beep two times

7. Finally, if you press the *x* or *X* key, the C++ program will be terminated

While compiling the code, you must also include the name of the wiringPi library, which is lwiringPi. The final compilation command looks as follows:

```
gcc -o LEDBuzzer -lncurses -lwiringPi LEDBuzzer.cpp
```

After compiling the code, type ./LEDBuzzer to run it:

Next, press the *L*, *E*, *D*, and *B* keys, and the LEDs and the buzzer will turn on and off accordingly.

Controlling a rover using a laptop keyboard

After controlling the LEDs and the buzzer, let's write a program for controlling our rover (the robot) from our laptop:

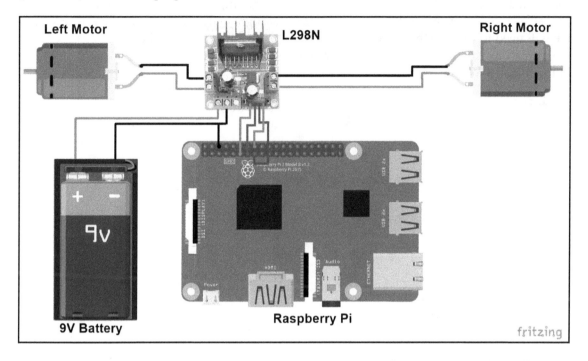

I have kept the wiring connections the same as they were in `Chapter 3`, *Programming the Robot*:

- The wiringPi pin numbers 0 and 2 are connected to the **IN1** and **IN2** pins
- The wiringPi pin numbers 3 and 4 are connected to the **IN3** and **IN4** pins
- The left motor pins are connected to the **OUT1** and **OUT2** pins of the motor driver
- The right motor pins are connected to the **OUT3** and **OUT4** pins of the motor driver
- Pin 6 of the Raspberry Pi is connected to the ground socket of the motor driver

Building a laptop-controlled rover program

If you have understood the previous two programs, by now, you might have figured out the code for our laptop-controlled rover. In this program, we will move the robot in forward, backward, left, and right directions using the up, down, left, and right arrow keys, as well as the *A*, *S*, *X*, *W*, and *D* keys. In order to recognize the inputs from the arrow keys, we will need to include the `keypad()` function inside our program. The `Laptop_Controlled_Rover.cpp` program can be download from the `Chapter05` folder of the `GitHub` repository:

```
int main()
{
...
for(;;)
{
initscr();
keypad(stdscr,TRUE);
refresh();
int keypressed = getch();
if(keypressed==KEY_UP || keypressed == 'W' || keypressed == 'w')
//KEY_UP command is for UP arrow key
{
printw("FORWARD");
digitalWrite(0,HIGH);
digitalWrite(2,LOW);
digitalWrite(3,HIGH);
digitalWrite(4,LOW);
}
if(keypressed==KEY_DOWN || keypressed == 'X' || keypressed == 'x')
//KEY_DOWN is for DOWN arrow key
{
printw("BACKWARD")
digitalWrite(0,LOW);
digitalWrite(2,HIGH);
digitalWrite(3,LOW);
digitalWrite(4,HIGH);
}

if(keypressed==KEY_LEFT || keypressed == 'A' || keypressed == 'a')
{
//KEY_LEFT is for LEFT arrow key
printw("LEFT TURN");
digitalWrite(0,LOW);
digitalWrite(2,HIGH);
digitalWrite(3,HIGH);
digitalWrite(4,LOW);
}
```

```
if(keypressed==KEY_RIGHT || keypressed == 'D' || keypressed == 'd')
{
//KEY_RIGHT is for right arrow keys
printw("RIGHT TURN");
digitalWrite(0,HIGH);
digitalWrite(2,LOW);
digitalWrite(3,LOW);
digitalWrite(4,HIGH);
}

if(keypressed=='S' || keypressed=='s')
{
printw("STOP");
digitalWrite(0,HIGH);
digitalWrite(2,HIGH);
digitalWrite(3,HIGH);
digitalWrite(4,HIGH);
}

if(keypressed=='E' || keypressed=='e')
{
break;
}
}
endwin();
return 0;
}
```

The preceding program can be explained as follows:

1. In the preceding program, if you press the up arrow key, this will be recognized by the KEY_UP code inside the first if condition. If the condition is TRUE, the robot will move forward, and the word FORWARD will be printed in the Terminal. Similarly, the robot will also move forward if you press the *W* or *w* keys.
2. If you press the down arrow key (KEY_DOWN) or the *X* or *x* keys, the robot will move backward, and the word BACKWARD will be printed in the Terminal.
3. If you press the left arrow key (KEY_LEFT) or the *A* or *a* keys, the robot will turn left, and the words LEFT TURN will be printed in the Terminal.
4. If you press the right arrow key (KEY_RIGHT) or the *D* or *d* keys, the robot will turn right, and the words RIGHT TURN will be printed in the Terminal.

5. Finally, if you press the *S* or *s* keys, the robot will stop, and the word STOP will be printed in the Terminal.
6. To terminate the code, we can press the *E* or *e* keys. Since we haven't provided any time delays, the robot will keep moving indefinitely, unless you stop the robot using the *S* or *s* keys.

When testing the code, connect the Raspberry Pi to a power bank so that your robot is completely wireless, and it can move freely.

Tracing a square path

After moving the robot in different directions, let's make the rover trace a square path. To do this, our robot will move as follows: forward -> right turn -> forward -> right turn -> forward -> right turn -> forward -> stop:

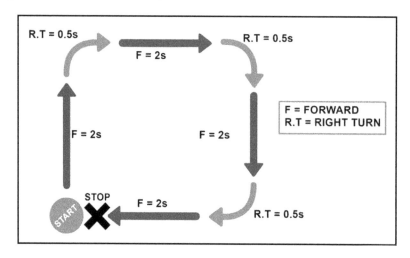

Inside the LaptopControlRover program, we will create another if condition. Inside this if condition, we will write a program to make the robot trace a square path. The if condition will look as follows:

```
if(keypressed == 'r' || keypressed == 'R')
{
forward(); //first forward movement
delay(2000);
rightturn(); //first left turn
delay(500); //delay needs to be such that the robot takes a perfect 90º
right turn
```

```
forward(); //second forward movement
delay(2000);
rightturn(); //second right turn
delay(500);

forward(); //third forward movement
delay(2000);
rightturn(); //third and last left turn
delay(500);

forward(); //fourth and last forward movement
delay(2000);
stop(); //stop condition
}
```

To trace the square path, the robot will move forward four times. It will take right turns three times, and, finally, it will stop. Outside the `main` function, we will need to create `forward()`, `rightturn()`, and `stop()` functions so that, instead of writing `digitalWrite` code multiple times inside the main function, we can simply call the necessary function.

Forward condition	Right turn	Stop

`void forward()`
`{`
`digitalWrite(0,HIGH);`
` digitalWrite(2,LOW);`
` digitalWrite(3,HIGH);`
` digitalWrite(4,LOW);`
`}`
	`void rightturn()`
	`{`
	`digitalWrite(0,HIGH);`
	` digitalWrite(2,LOW);`
	` digitalWrite(3,LOW);`
	` digitalWrite(4,HIGH);`
	`}`
		`void stop()`
		`{`
		`digitalWrite(0,HIGH);`
		` digitalWrite(2,HIGH);`
		` digitalWrite(3,HIGH);`
		` digitalWrite(4,HIGH);`
		`}`

This is how we can control the robot using a laptop, with the help of the keyboard keys. Next, let's take a look at the second technique, in which we will create GUI buttons using QT5. When these buttons are pressed, the robot will move in different directions.

Installing and setting up QT5

QT is a cross-platform application framework generally used for embedded graphical user interfaces. The latest version of QT is 5, so it is also referred to as QT5. To install the QT5 software inside our RPi, open the Terminal window and type in the following command:

```
sudo apt-get install qt5-default
```

The output of the preceding command is shown in the following screenshot:

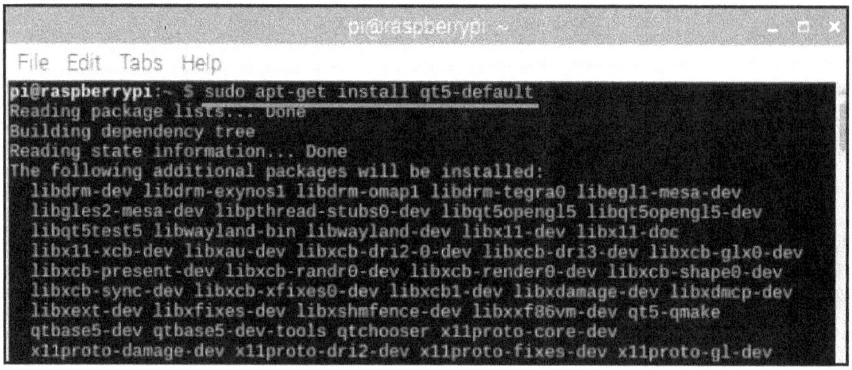

This command will download the necessary qt5 files that run in the backend. Next, for downloading and installing the QT5 IDE, type in the following command:

```
sudo apt-get install qtcreator
```

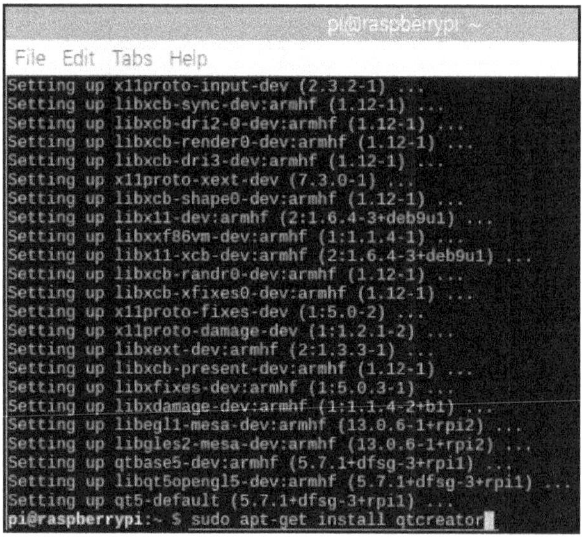

The installation of QT5 IDE will take around 10 to 15 minutes depending on your internet speed. If you face any problems while installing QT5, try updating and upgrading your RPi. To do this, type the following commands in your Terminal window:

```
sudo apt-get update
sudo apt-get upgrade -y
```

Setting up QT5

Before we write any programs inside QT5, we first need to set it up so that it can run C++ programs. To open QT5, click on the raspberry icon, go to **Programming**, then select the **Qt Creator**:

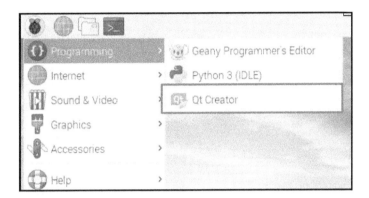

QT5 runs a bit slower in RPi, so it will take some time for the IDE to open. Click on **Tools** and then select **Options...**:

Inside **Options...**, click on **Devices** and make sure the **Type** is set to **Desktop**. The name should be `Local PC`, which refers to the RPi:

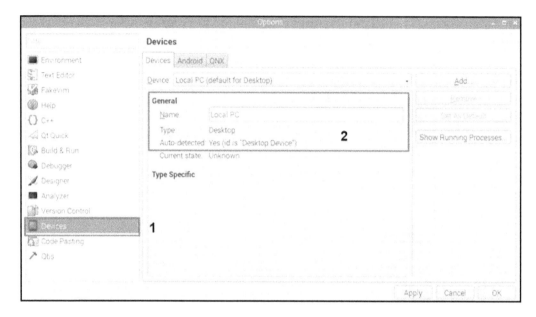

After that, click on the **Build & Run** option. Next, select the **Kits** tab and click on the **Desktop (default)** option:

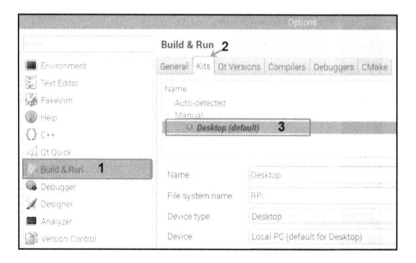

After selecting the **Build & Run** option, there are a couple of modifications that we have to make:

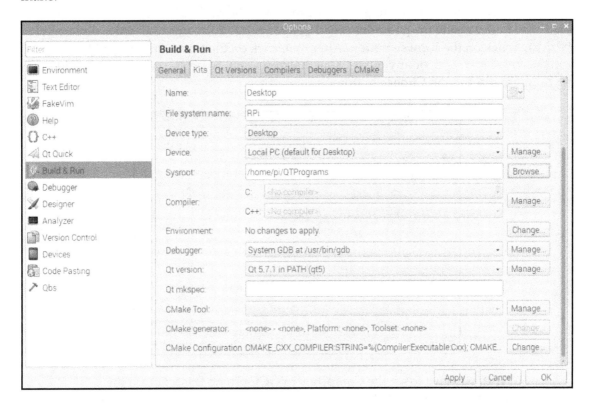

Let's see the modifications step by step:

1. Keep the **Name** as `Desktop`.
2. Set the name of the filesystem to `RPi`.
3. In **Device type**, select the **Desktop** option.

4. The **Sysroot** (system root) will be set to /home/pi by default, which means that, when we create a new QT5 application, it will be created inside the pi folder. Now, instead of creating our QT projects in the pi folder, we will create a new folder inside the pi folder called QTPrograms. To change the folder directory, click on the **Browse** button. After that, click on the folder option. Call this folder QTPrograms, or any other name that you want. Select the QTPrograms folder and select the **Choose** button:

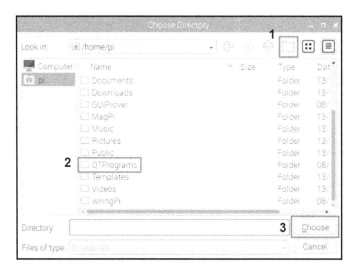

5. Next, we have to set the **Compilers** to **GCC**. To do this, click on the **Compilers** tab. Inside this, click on the **Add** drop-down button. Go to **GCC** and select the **C++** option:

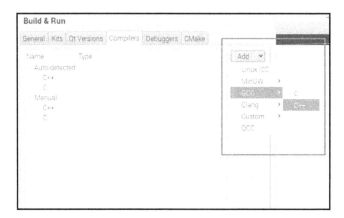

You will now see the **GCC** compilation option below the **C++** option:

After this, click on the **Apply** button to apply the changes and then click on the **OK** button. Next, click on **Tools** again and open **Options**. Inside the Build and run option, select the **Kits** tab and, again, select the **Desktop** option. This time, next to the **C++** option, you will see a drop-down option. Click on this, and select the **GCC** compiler:

6. Next, check the **Debugger** option. It should be set to **System GDB at /usr/bin/gdb**.

7. Finally, check the QT5 version. Currently, I'm using the latest version of QT, which is 5.7.1. By the time you come across this chapter, the latest version is likely to have been updated.

After making these changes, press **Apply** and then **OK**. After setting up the QT5, let's write our first program to turn an LED on and off using the GUI buttons.

Controlling LEDs with GUI buttons

In this section, we will create a simple QT5 program in which we will turn the LEDs on and off using the GUI buttons. For this project, you will need two LEDs:

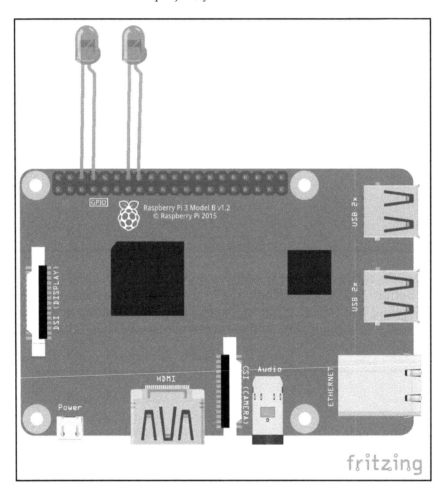

The wiring connections of the LEDs is exactly the same as that in the LEDBuzzer project:

- The anode (positive) pin of the first LED is connected to the wiringPi pin number 0 and the cathode (negative) pin is connected the physical pin number 9 (the ground pin)
- The anode pin of the second LED is connected to the wiringPi pin number 2 and the cathode pin is connected to the physical pin number 14 (the ground pin)

Creating a QT project

The QT5 project for turning the LEDs on and off is called LedOnOff. You can download this project from the Chapter05 folder of the GitHub repository. After downloading the LedOnOff project folder, open the LedOnOff.pro file to view the project inside the QT5 IDE.

Follow these steps to create a project in the QT5 IDE:

1. click on the **File** option and then click on **New File or Project...**:

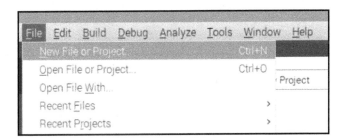

2. Next, select the **QT Widgets Application** option and click the **Choose** button:

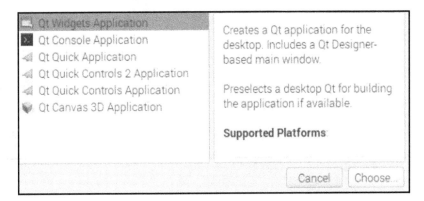

3. After that, give your project a name. I have named my project LEDOnOff. After this, change the directory to QTPrograms so that the project is created in this folder, then click **Next**:

4. Keep the **Desktop** option checked and then click **Next**:

5. You should now see certain filenames, which are a part of this project. Keep the names as they are and click **Next**:

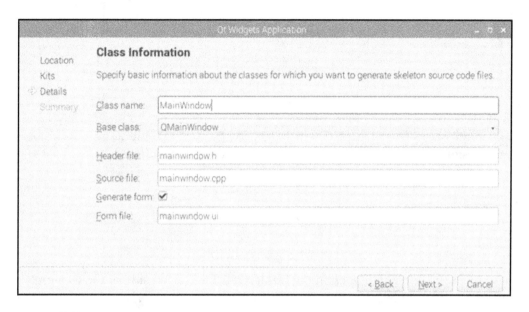

6. Finally, you will see a **Summary** window, which will show you a summary of all the files that will be created. We don't have to make any changes in this window, so click on **Finish** to create the project:

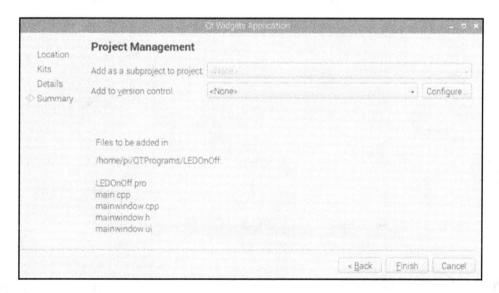

On the left side of the IDE, you will see the design, C++, and header files. First, we will open the `LEDOnOff.pro` file and add the path of the `wiringPi` library. At the bottom of this file, add the following code:

```
LIBS += -L/usr/local/lib -lwiringPi
```

Next, open the `mainwindow.ui` file, which is inside the `Forms` folder. The `mainwindow.ui` file is the designer file inside of which we will design our GUI buttons:

The `mainwindow.ui` file will open in the **Design** tab. On the left side of the **Design** tab is the widget box, which contains widgets such as buttons, a list view, and layouts. In the middle is the design area, where we will drag the UI components. In the bottom-right, the properties of the selected UI components are displayed:

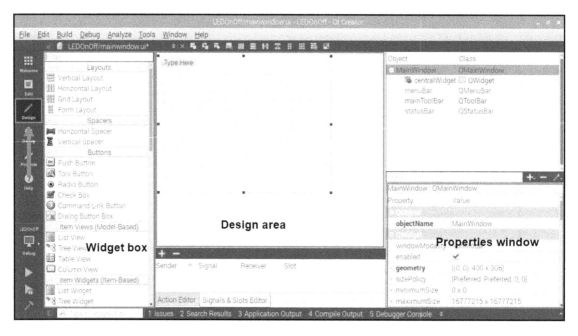

Next, to create the GUI button, drag the **Push Button** widget inside the design area. Double-click on the button, and change the text to ON. After that, with the **Push Button** selected, change the **objectName** (inside the properties window) to on:

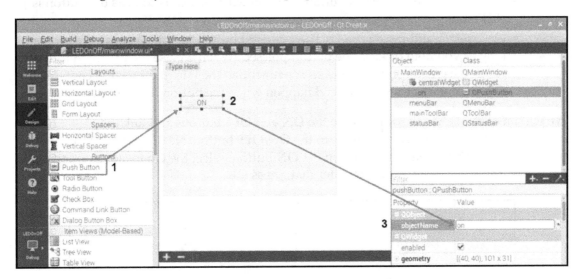

After this, add two more buttons. Set the name of one button to OFF and the **objectName** to off. Set the name of another button to ON / OFF and the **objectName** to onoff:

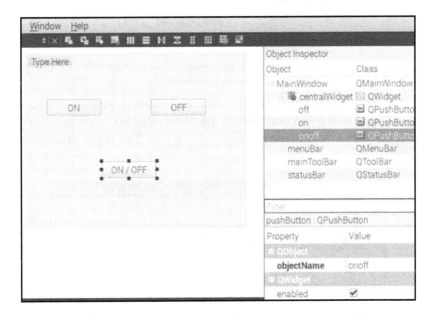

There are two different types of button function that we can use for turning the LED on and off:

- `clicked()`: The `clicked` button function will execute as soon as the button is clicked.
- `pressed()` and `released()`: The `pressed` button function keeps executing as long as you hold or keep the button pressed. When we use the `pressed` function, we also have to use the `released()` function. The released function contains the code that indicates what should happen when the button is released.

We will link the `clicked()` function to the **ON** and **OFF** buttons and link the `pressed()` and `released()` functions to the **ON/OFF** button. Next, to link the `clicked()` function to the **ON** button, right-click on the **ON** button, select the **Go to slot...** option, and then select the `clicked()` function. After that, press **OK**:

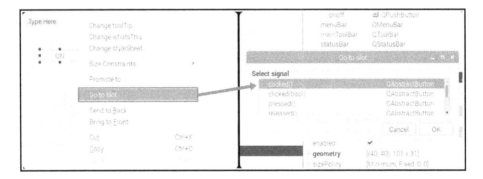

Now, as soon as you select the `clicked()` function, a clicked function called `on_on_clicked()` (`on_buttonsobjectname_clicked`) will be created inside the `mainwindow.cpp` file (this file is inside the `Sources` folder). Inside this function, we will write the program to turn the LED on. Before that, however, we need to declare the `wiringPi` library and pins inside the `mainwindow.h` file. This file is inside the `Headers` folder:

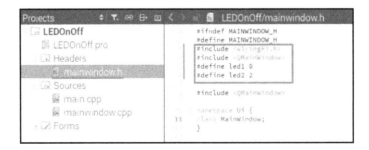

We also need to declare the `QMainWindow` library, which will create a window that contains our buttons. Next, I have set the `led1` pin to pin `0` and the `led2` pin to pin `2`. After that, open the `mainwindow.cpp` file again. We will then do the following:

1. First, we will declare the `wiringPiSetup();` function
2. Next, we will set `led1` and `led2` as the `OUTPUT` pins
3. Finally, inside the `on_on_clicked()` function, set the `led1` and `led2` pins to `HIGH`:

Next, to turn the LEDs off, open the `mainwindow.ui` file again, right-click on the off button, select **Go to slot...**, and select the `clicked()` function again. Inside the `mainwindow.cpp` file, a new function called `on_off_clicked` will be created. Inside this function, we will write the program to turn off the LEDs.

To program the ON/OFF button, right-click on it, select **Go to slot...**, and, this time, select the `pressed()` function. A new function name of `on_onoff_pressed()` will be created inside the `mainwindow.ui` file. Next, right-click on the **ON/OFF** button, select **Go to slot...**, and select the `released()` function. A new function called `on _onoff_released()` will now be created.

Inside the `on_onoff_pressed()` function, we will write a program to turn the LEDs on. Inside the `on_onoff_released()` function, we will write the program to turn the LEDs off:

```
void MainWindow::on_onoff_pressed()
{
    digitalWrite(led1,HIGH);
    digitalWrite(led2,HIGH);
}

void MainWindow::on_onoff_released()
{
    digitalWrite(led1,LOW);
    digitalWrite(led2,LOW);
}
```

Before running the code, click on **File** and then click **Save All**. Next, to build and run the code, click on **Build** and then click on the **Run** option. It will take around 30 to 40 seconds for the **MainWindow** to appear, and, in the main window, you will see the GUI buttons as follows:

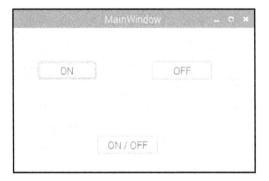

Now, when you click the **ON** button, the LEDs will turn on. When you click the **OFF** button, the LEDs will turn off. Finally, when you hold the **ON / OFF** button, the LEDs will turn on until you let go, when they will turn off.

Dealing with errors

In the console, you may see some minor errors. If the main window is open, you can ignore these errors:

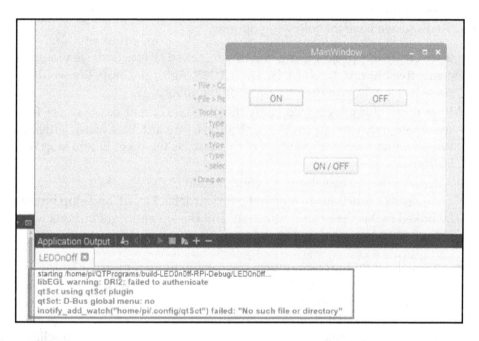

The GCC compiler might keep resetting when you open the **Qt Creator** IDE. Because of this, after running the project, you will get the following error:

```
Error while building/deploying project LEDOnOff (kit: Desktop)
  When executing step "qmake"
```

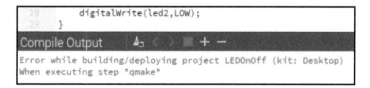

If you get this error, go to **Tools**, then **Options**, and set the **C++** compiler to **GCC**, as indicated in *step 5* of the *Setting up QT5* section.

Controlling a rover using a laptop with QT5

Now that we can control the LEDs, let's look at how to control the rover using QT5. Inside the **Qt Creator** IDE, create a new project and name it QTRover. You can download the QTRover project folder from the GitHub repository of this chapter. We can now create this QTRover project using the clicked() function and the pressed() and released() functions. To do so, we have the following options:

1. If we create this project using only the clicked() function, we would need to create five buttons: forward, backward, left, right, and stop. We would need to press the stop button each time to stop the robot.
2. If we create this project using only the pressed() and released() functions, we would only need to create four buttons: forward, backward, left, and right. We wouldn't need a stop button in this case, as the rover would stop when the buttons are released.
3. Alternatively, we can also use a combination of the clicked(), pressed(), and released() functions in which the forward, backward, and stop buttons would be linked to the clicked() function, and the left and right buttons would be linked to the pressed() and released() functions.

In this project, we'll opt for the third option, the combination of the clicked(), pressed(), and released() functions. Before creating this project, we will close the LEDOnOff project, because if both the LEDOnOff and QTRover projects are kept open, there is a chance that, if you make UI changes in one project, the code might change in the other, thus affecting both of your project files. To close the LEDOnOff project, right-click on it and then select the **Close Project "LEDOnOff"** option.

Next, add the wiringPi library path inside the QTRover.pro file:

```
LIBS += -L/usr/local/lib -lwiringPi
```

After that, open the `mainwindow.ui` file and create five push buttons. Label them FORWARD, BACKWARD, LEFT, RIGHT, and STOP:

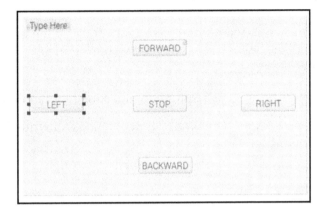

Set the name of the button objects as follows:

- Set the FORWARD button object name to forward
- Set the BACKWARD button object name to backward
- Set the LEFT button object name to left
- Set the RIGHT button object name to right
- Set the STOP button object name to stop

After this, right-click on the forward, backward, and stop buttons, and add the `clicked()` functions to those three buttons. Similarly, right-click on the left and right buttons, and add the `pressed()` and `released()` functions to these buttons.

Next, open the `mainwindow.h` file and declare the `wiringPi` and `QMainWindow` libraries. Also, declare the four `wiringPi` pin numbers. In my case, I'm using the pin numbers 0, 2, 3, and 4:

Inside the `mainwindow.cpp` file, we will have three `on_click` functions to move forward (`on_forward_clicked`), backward (`on_backward_clicked`), and to stop (`on_stop_clicked`).

We also have two `on_pressed` and `on_released` functions for the left (`on_left_pressed` and `on_left_released`) and right (`on_right_pressed` and `on_right_released`) buttons.

The following steps describe the steps required to move the robot in different directions:

1. Inside the `on_forward_clicked()` function, we will write the program to move the robot forward:

   ```
   digitalWrite(leftmotor1, HIGH);
   digitalWrite(leftmotor2, LOW);
   digitalWrite(rightmotor1, HIGH);
   digitalWrite(rightmotor2, LOW);
   ```

2. Next, inside the `on_backward_clicked()` function, we will write the program to move the robot backward:

   ```
   digitalWrite(leftmotor1, HIGH);
   digitalWrite(leftmotor2, LOW);
   digitalWrite(rightmotor1, HIGH);
   digitalWrite(rightmotor2, LOW);
   ```

3. After that, inside the `on_left_pressed()` function, we will write the program to make an axial left turn or radial left turn:

   ```
   digitalWrite(leftmotor1, LOW);
   digitalWrite(leftmotor2, HIGH);
   digitalWrite(rightmotor1, HIGH);
   digitalWrite(rightmotor2, LOW);
   ```

4. Then, inside the `on_right_pressed()` function, we will write the program to make an axial right turn, or a radial right turn:

   ```
   digitalWrite(leftmotor1, HIGH);
   digitalWrite(leftmotor2, LOW);
   digitalWrite(rightmotor1, LOW);
   digitalWrite(rightmotor2, HIGH);
   ```

5. Inside the `on_stop_clicked()` functions, we will write the program to stop the robot:

```
digitalWrite(leftmotor1, HIGH);
digitalWrite(leftmotor2, HIGH);
digitalWrite(rightmotor1, HIGH);
digitalWrite(rightmotor2, HIGH);
```

After completing the code, save all the files. After that, run the program and test the final output. After running the code, you will see the main window with the forward, backward, left, right, and stop buttons. Press each GUI button to move the robot in the desired direction.

Summary

In this chapter, we looked at two different techniques for controlling the robot using a laptop. In the first technique, we used the `ncurses` library to take input from the keyboard to move the robot accordingly. In the second technique, we used the **QT Creator** IDE to create GUI buttons, and then used these to move the robot in different directions.

In the next chapter, we will install OpenCV software on our Raspberry Pi. After that, we will use the Raspberry Pi camera to record pictures and videos.

Questions

1. The `ncurses` program should be written between which two functions?

2. What is the purpose of the `initscr()` function?

3. How do you compile a `ncurses` code inside the Terminal window?

4. Which C++ compiler did we use in the QT Creator?

5. Which push-button function, or functions, will you use to move the robot forward as long as the button is pressed?

Section 3: Face and Object Recognition Robot

In this section you will use OpenCV to detect faces and real-world objects. We will then extend the power of OpenCV to recognize different faces and move the robot once the correct face is detected.

The following chapters are included in this section:

- Chapter 6, *Accessing the RPi Camera with OpenCV*
- Chapter 7, *Building an Object-Following Robot with OpenCV*
- Chapter 8, *Face Detection and Tracking Using the Haar Classifier*

6
Accessing the RPi Camera with OpenCV

One of the most exciting things we can do with Raspberry Pi is recognize objects and faces by connecting it to an external USB webcam or **Raspberry Pi camera** (**RPi camera**).

To process these inputs from the camera, we will use the OpenCV libraries. As the installation of OpenCV takes a long time and involves multiple steps, this whole chapter will be dedicated to getting you up and running.

In this chapter, you will explore the following topics:

- Installing OpenCV 4.0.0 on Raspberry Pi
- Enabling and connecting the RPi camera to RPi
- Capturing images and video with the RPi camera
- Reading an image using OpenCV

Technical requirements

In this chapter, you will need the following:

- A Raspberry Pi camera module – as of 2019, the latest RPi camera module is called **RPi Camera V2 1080P**
- A Raspberry Pi camera case (mount)

The code file for this chapter can be downloaded from `https://github.com/ PacktPublishing/Hands-On-Robotics-Programming-with-Cpp/tree/master/Chapter06`.

Installing OpenCV 4.0.0 on Raspberry Pi

The **Open Source Computer Vision Library** (**OpenCV**) is an open source computer vision and machine learning library. The OpenCV library consists of more than 2,500 computer vision and machine learning algorithms that can be used to recognize objects, detect colors, and track moving objects in real life or in a video. OpenCV supports the C++, Python, and Java programming languages, and can run on Windows, macOS, Android, and Linux.

Installing OpenCV on a Raspberry Pi is a time-consuming and lengthy process. There are multiple libraries and files that we have to install along with the OpenCV library to make it work properly. The steps to install OpenCV will be performed on my Raspberry Pi 3B+ model, which is running Raspbian Stretch. The OpenCV version we are going to install is OpenCV 4.0.0.

 While installing OpenCV, we will download multiple files. If you live in a big house, make sure that you sit near the Wi-Fi router so that the RPi receives good signal strength. If the RPi is far away from the Wi-Fi, the download speed may be affected and it may take you more time to install OpenCV on your RPi. It took me around 3 hours to install OpenCV on my RPi 3B+ with a download speed of around 500-560 Kbps.

Uninstalling Wolfram and LibreOffice

If you are using a 32 GB microSD card, Raspbian Stretch will occupy only 15% of the storage space, but if you are using an 8 GB microSD card, it will occupy 50% of the space. If you are using an 8 GB microSD card, you will need to free up some space. You can do this by uninstalling some unused apps. Two such apps are Wolfram engine and LibreOffice.

Uninstalling apps on Raspbian Stretch is easy. You just need to enter a command in a Terminal window. Let's begin by uninstalling the Wolfram engine:

```
sudo apt-get purge wolfram-engine -y
```

Next, use the same command to uninstall LibreOffice:

```
sudo apt-get purge libreoffice* -y
```

After uninstalling both pieces of software, let's do some cleaning by using two simple commands:

```
sudo apt-get clean
sudo apt-get autoremove -y
```

Now that we have freed up some space, let's update our RPi.

Updating your RPi

Updating your RPi involves a few simple steps:

1. Open a Terminal window and enter the following command:

    ```
    sudo apt-get update
    ```

2. Upgrade the RPi by entering the following command:

    ```
    sudo apt-get upgrade -y
    ```

3. Reboot the RPi:

    ```
    sudo shutdown -r now
    ```

Once your RPi restarts, open the Terminal window again.

 While running certain commands in the Terminal window, you may get a prompt that asks you whether you want to continue. In the commands for this process, we have already added the -y command (at the end of the line), which will automatically apply the **yes** command to the prompts.

Installing the cmake, image, video, and gtk packages

cmake is a configuration utility. Using cmake, we can configure different OpenCV and Python modules after installing them. To install the cmake package, enter the following command:

```
sudo apt-get install build-essential cmake pkg-config -y
```

Next, to install the image I/O package, enter the following command:

```
sudo apt-get install libjpeg-dev libtiff5-dev libjasper-dev libpng12-dev -y
```

After this, we will install the two video I/O packages by typing the following commands:

```
sudo apt-get install libavcodec-dev libavformat-dev libswscale-dev libv4l-
dev -y
sudo apt-get install libxvidcore-dev libx264-dev -y
```

Next, we will download and install the **Gimp Toolkit** (**GTK**) packages. This toolkit is used for making graphical interfaces for our program. We will execute the following commands to download and install the GTK packages:

```
sudo apt-get install libgtk2.0-dev libgtk-3-dev -y
sudo apt-get install libatlas-base-dev gfortran -y
```

Downloading and unzipping OpenCV 4.0 and its contribution repository

Once we have installed these packages, we can move on to OpenCV. Let's begin by downloading Open CV 4.0:

1. Type the following command into your Terminal window:

   ```
   wget -O opencv.zip
   https://github.com/opencv/opencv/archive/4.0.0.zip
   ```

2. Download the `opencv_contrib` repository, which contains some additional modules. Enter the following command:

   ```
   wget -O opencv_contrib.zip
   https://github.com/opencv/opencv_contrib/archive/4.0.0.zip
   ```

   The commands in *step 1* and *step 2* are both single-line commands.

3. Unzip the `opencv.zip` file using the following command:

   ```
   unzip opencv.zip
   ```

4. Unzip the `opencv_contrib.zip` file:

   ```
   unzip opencv_contrib.zip
   ```

After unzipping `opencv` and `opencv_contrib`, you should see the `opencv-4.0.0` and `opencv_contrib-4.0.0` folders inside the `pi` folder.

Installing Python

Next, we will install Python 3 and some of its support tools. Even though we are going to program in OpenCV using C++, it is still a good idea to install and link Python packages with OpenCV so that you have the option to write or compile Python codes using OpenCV.

To install Python and its developments tools, type in the following commands:

```
sudo apt-get install python3 python3-setuptools python3-dev -y
wget https://bootstrap.pypa.io/get-pip.py
sudo python3 get-pip.py
sudo pip3 install numpy
```

After installing the Python packages, we can compile and build OpenCV.

Compiling and installing OpenCV

To compile and install OpenCV, we need to go through the following steps:

1. Go inside the `opencv-4.0.0` folder. Change the directory to the `opencv-4.0.0` folder using the following command:

   ```
   cd opencv-4.0.0
   ```

2. Inside this folder, create a `build` folder. To do so, type the following command:

   ```
   mkdir build
   ```

3. To open the `build` directory, type the following command:

   ```
   cd build
   ```

4. After changing the directory to `build`, enter the following command:

   ```
   cmake -D CMAKE_BUILD_TYPE=RELEASE \
   -D CMAKE_INSTALL_PREFIX=/usr/local \
   -D BUILD_opencv_java=OFF \
   -D BUILD_opencv_python2=OFF \
   -D BUILD_opencv_python3=ON \
   -D PYTHON_DEFAULT_EXECUTABLE=$(which python3) \
   -D INSTALL_C_EXAMPLES=ON \
   -D INSTALL_PYTHON_EXAMPLES=ON \
   -D BUILD_EXAMPLES=ON\
   -D OPENCV_EXTRA_MODULES_PATH=~/opencv_contrib-4.0.0/modules \
   -D WITH_CUDA=OFF \
   ```

```
-D BUILD_TESTS=OFF \
-D BUILD_PERF_TESTS= OFF ..
```

Make sure that you copy the two dots, . ., at the end when entering this command in the Terminal window.

5. To enable all four cores of the RPi, open the `swapfile` file inside the nano editor:

```
sudo nano /etc/dphys-swapfile
```

6. Inside this file, search for the `CONF_SWAPSIZE=100` code and change the value from `100` to `1024`:

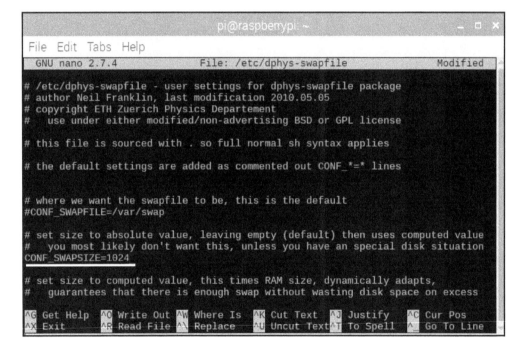

7. Press *Ctrl + O* to save this file. You will get a prompt at the bottom of the file, asking you whether you want to save this file. Press *Enter* and then press *Ctrl + X* to exit.

8. To apply these changes, type in the following two commands:

```
sudo /etc/init.d/dphys-swapfile stop
sudo /etc/init.d/dphys-swapfile start
```

9. To compile OpenCV with all four cores of the RPi, type in the following command:

```
make -j4
```

This is the most time-consuming step and it will take anywhere between 1.5 to 2 hours. If you face any errors while compiling, try compiling with one core.

To compile with only one core, enter the following command:

```
sudo make install
make
```

Use the preceding two commands only if you encounter an error with the make -j4 command.

10. To install OpenCV 4.0.0, enter the following commands:

```
sudo make install
sudo ldconfig
```

We've now compiled and installed OpenCV. Let's connect it to Python.

Linking OpenCV to Python

Let's follow these steps to link OpenCV to Python:

1. Open the python 3.5 folder (/usr/local/python/cv2/python-3.5):

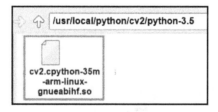

Inside this folder, you should see a file named cv2.so or cv2.cpython-35m-arm-linux-gnueabihf.so. If the filename is cv2.so, you do not need to make any changes. If the filename is cv2.cpython-35m-arm-linux-gnueabihf.so, you have to rename it to cv2.so. To rename this file, change the directory to python 3.5 by entering the following command:

```
cd /usr/local/python/cv2/python-3.5
```

To rename this file from `cv2.cpython-35m-arm-linux-gnueabihf.so` to `cv2.so`, enter the following command:

```
sudo mv /usr/local/python/cv2/python3.5/cv2.cpython-35m-arm-linux-gnueabihf.so cv2.so
```

2. Move this file to the `dist-package` folder (`/usr/local/lib/python3.5/dist-packages/`) using the following command:

```
sudo mv /usr/local/python/cv2/python-3.5/cv2.so
/usr/local/lib/python3.5/dist-packages/cv2.so
```

3. To test whether OpenCV 4.0.0 is linked properly to Python 3, go to the `pi` directory by typing `cd ~` in the Terminal window. Next, type `python3`:

4. You should see a three-corner bracket. Type `import cv2`.

5. To check the OpenCV version, type `cv2.__version__`. If you see `opencv 4.0.0`, this means that OpenCV is successfully installed and linked with the Python packages:

6. Type exit() and press *Enter*:

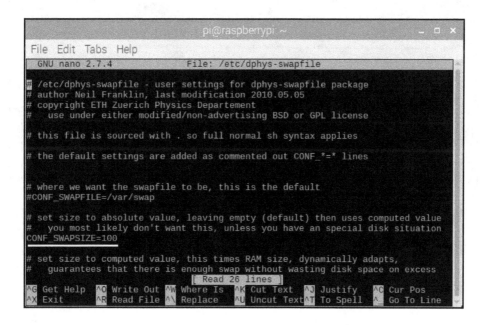

After installing OpenCV, we need to reset CONF_SWAPSIZE back to 100:

1. Open swapfile:

   sudo nano /etc/dphys-swapfile

2. Change CONF_SWAPSIZE to 100:

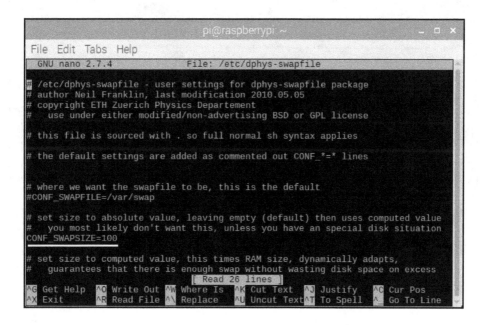

3. To apply these changes, enter the following commands:

   sudo /etc/init.d/dphys-swapfile stop
   sudo /etc/init.d/dphys-swapfile start

You have successfully installed OpenCV 4.0.0 on your Raspberry Pi. We are now ready to attach the RPi camera to the RPi.

Enabling and connecting the RPi camera to RPi

Before connecting the RPi camera to the RPi, we need to enable the **Camera** option from the RPi configuration:

1. Open a Terminal window and enter `sudo raspi-config` to open the RPi configuration.
2. Select **Advanced Options** and press *Enter* to open it:

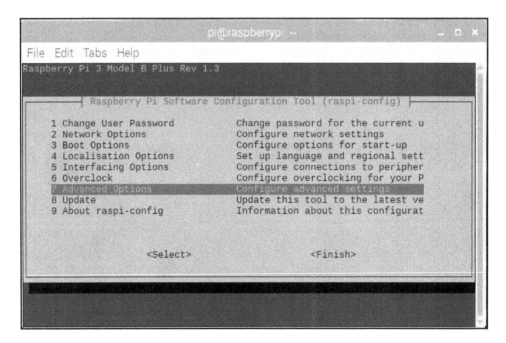

3. Select the **Camera** option and press *Enter* to open it:

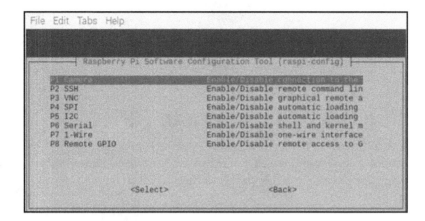

4. Select **Yes** and press *Enter* to enable the **Camera** option:

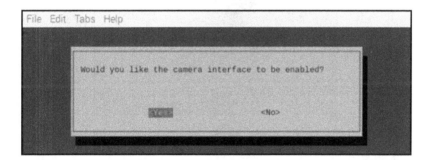

5. Select **Ok** and press *Enter*:

6. Exit the RPi configuration and shut down your RPi.

While connecting the RPi camera to the RPi, make sure that the RPi is turned off.

Now that we have finished the setup, let's connect the camera.

Connecting the RPi camera to RPi

Connecting the RPi camera to RPi is an easy, but delicate, process. The RPi camera has a ribbon wire connected to it. We have to insert this ribbon wire inside the camera slot of the RPi, which is placed between the LAN port and the HDMI port:

The ribbon on the RPi camera consists of a blue strip on the front and nothing on the back:

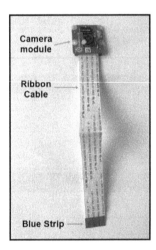

Now that we understand the component and the port, let's start connecting them:

1. Gently lift up the lid of the camera slot:

2. Insert the camera ribbon inside the slot, making sure that the blue tape on the ribbon faces the LAN port.
3. Press the lid to lock the camera ribbon:

That's it—your RPi camera is now ready to take pictures and record videos.

Mounting the RPi camera on the robot

Let's mount the RPi camera on the robot; you will need a RPi camera case for this. A quick search on `amazon.com` for `RPi camera case` will show the following case:

I do not recommend this particular case as it did not fit my RPi camera module properly. When the case is closed, the lens of my RPi camera did not properly align with the small hole of this camera case.

Since I live in India, I did not find any good RPi camera cases on the Amazon India website (`www.amazon.in`), and the ones that were available were expensive. The case I ended up using was from an Indian e-commerce website called `www.robu.in`, and it cost me only 90 Rs (less than $2). Before purchasing a camera case or camera mount from an e-commerce website, please check the reviews to ensure that it won't damage your RPi camera.

The image of the RPi camera case that I used is shown in the following image. I bought this case from an Indian website called `www.robu.in`. On this website, search for `Camera mount module for Raspberry Pi` to find this camera mount:

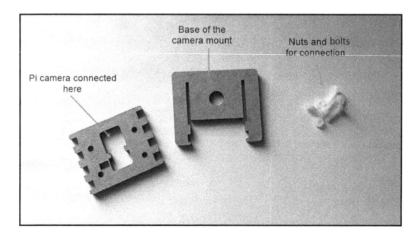

Even though this camera mount contains four small nuts and bolts to attach the RPi camera to the camera mount, I found that the threading of the nut and bolt was inaccurate, and attaching the RPi camera to the camera mount was way too difficult. Therefore, I used four small pieces of double-sided tape and attached the tape to the holes of the RPi camera:

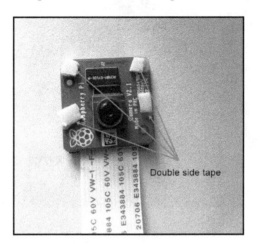

Next, I attached the RPi camera to the camera mount. In the following image, the RPi camera is fitted upside down. So, when we capture an image, the image will appear upside down, and to view the image properly, we need to flip it (the process of flipping the image horizontally and vertically inside OpenCV is explained in Chapter 7, *Building an Object-Following Robot with OpenCV*):

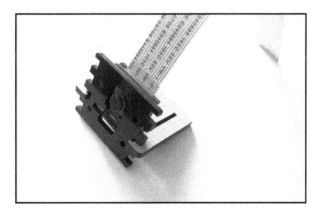

After this, I attached the camera mount on top of my RPi case using small strips of double-sided tape, thus mounting the RPi camera on the robot:

Now that we've mounted the camera case on the robot, let's see how we can capture images and video using the RPi camera.

Capturing images and video with the RPi camera

Let's see how we can take pictures and record video with our RPi. Open the Terminal window and type the following command:

```
raspistill -o image1.jpg
```

In this command, we used `raspistill` to take a still picture and saved it as `image1.jpg`.

Since the Terminal window is pointing to the `pi` directory, this image is saved in the `pi` folder. To open this image, open the `pi` folder and inside it, you will see `image1.jpg`. Images captured using the RPi camera have a native resolution of 3,280 x 2,464 pixels:

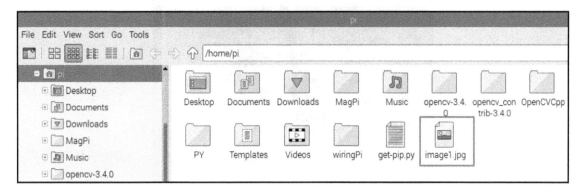

The output of `image1` is shown in the following screenshot:

If we want to flip the image horizontally, we can add the `-hf` command, and if we want to flip it vertically, we can add `-vf` command inside the `raspistill` code:

```
raspistill -hf -vf -o image2.jpg
```

The `image2.jpg` file is also saved in the `pi` folder, and its output is shown in the following screenshot:

Now that we have captured images using the RPi camera, let's record and view a video.

Recording a video with the RPi camera

Now that we know how to take a picture with the RPi camera, let's look at how to record video with it. The command for recording a video clip is as follows:

```
raspivid -o video1.h264 -t 5000
```

The preceding command doesn't yield any output, as shown in the following screenshot:

In our command, we used `raspivid` to record the video and named it `video1`. We recorded the video in the `h264` format. The number `5000` represents 5,000 milliseconds, that is, we recorded a 5-second video. You can open the `pi` folder and double-click on the video file to open it:

Now that we know how to take a picture and record a video, let's install the `v4l2` driver so that the OpenCV libraries can detect the RPi camera.

Installing the v4l2 driver

OpenCV libraries can, by default, recognize the USB camera attached to the RPi's USB port, but it cannot directly detect the RPi camera. To recognize our RPi camera, we need to load the `v4l2` driver inside the modules file. To open this file, enter the following command inside the Terminal window:

```
sudo nano /etc/modules
```

To load the `v4l2` driver, add `bcm2835-v4l2` inside the following file:

```
                          pi@raspberrypi: ~                        _  □  ✕
File  Edit  Tabs  Help
  GNU nano 2.7.4                    File: /etc/modules
# /etc/modules: kernel modules to load at boot time.
#
# This file contains the names of kernel modules that should be loaded
# at boot time, one per line. Lines beginning with "#" are ignored.

i2c-dev
bcm2835-v4l2
```

Press *Ctrl + O*, followed by *Enter*, to save this file and press *Ctrl + X* to exit the file, then reboot your RPi. Once rebooted, the RPi camera will be recognized by the OpenCV libraries.

Reading an image using OpenCV

After playing around a little bit with the RPi camera, let's write a simple C++ program using the OpenCV functions to display an image. In this program, we first read an image from a particular folder and then we display this image in a new window:

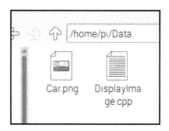

To display the image, we first need an image. Inside the `pi` folder, I have created a new folder called `Data` and inside it, I have copied one image called `Car.png`. In the same folder, I have created the `DisplayImage.cpp` file, inside which we are going to write the program to display the image. The `DisplayImage.cpp` program can be download from the `Chapter06` folder of this book's GitHub repository. The code is as follows:

```
#include <iostream>
#include <stdio.h>
#include <opencv2/opencv.hpp>

using namespace cv;
using namespace std;
int main()
{

Mat img;

img = imread("Car.jpg");

imshow("Car Image", img);

waitKey(0);

return 0;
}
```

In the preceding code, we first declared the opencv.hpp library, along with the basic C++ libraries. We then declared the cv namespace, which is a part of the OpenCV library. After this, inside the main function, we declared a matrix (Mat) variable called img.

Next, the imread() function is used to read the Car.jpg image and the value is stored in the img variable. If the image and .cpp file are in the same folder, you just write the image name inside the imread() function. If the image is in a different folder, the location of the image should be mentioned inside the imread function.

The imshow() function is used to display the car image in a new window. The imshow() function takes two parameters as input. The first parameter is the window text ("Car Image") and the second parameter is the variable name (img) of the image that is to be displayed.

The waitKey(0) function is used to create an infinite delay, that is, the waitKey(0) will display the car image infinitely until you press any key. When a key is pressed, the next set of code will execute. Since we do not have any code after the waitKey(0) function, the program will terminate and the car image window will be closed.

To compile and build OpenCV codes inside RPi, we need to add the following lines inside the compilation and build box:

1. Click on **Build option** and then select **Set Build Commands**. Inside the compilation box, enter the following commands:

```
g++ -Wall $(pkg-config --cflags opencv) -c "%f" -lwiringPi
```

2. Inside the **Build** box, enter the following command and then click **OK**:

```
g++ -Wall $(pkg-config --libs opencv) -o "%e" "%f" -lwiringPi
```

3. Click on the compile button to compile the code and then click on the build button to test the output. In the output, a new window will be created, inside which the car image will be displayed:

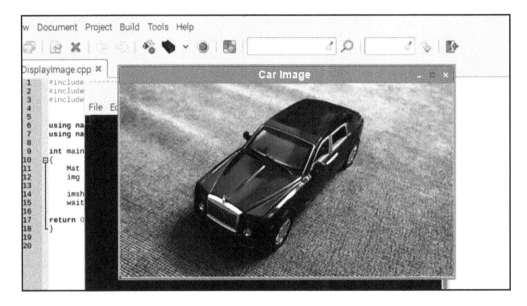

4. If you press any key, the program will terminate and the car image window will be closed.

Summary

In this chapter, we focused on installing OpenCV on Raspberry Pi. You were introduced to the RPi camera module. After setting up the RPi camera, you took pictures and recorded a short video clip using the RPi camera.

In the next chapter, we are going to write C++ programs using the OpenCV libraries. You will learn about different image-processing concepts so that you can scan, threshold, and recognize an object. After recognizing an object, we will write programs for the robot so that it follows the object.

Questions

1. What is the full form of OpenCV?
2. What is the resolution of images that are captured by the RPi camera?
3. What is the command to capture an image using the RPi camera?
4. What is the command to record a video using the RPi camera?
5. What percentage of the memory does Raspbian OS occupy on an 8 GB and 32 GB SD card?

7
Building an Object-Following Robot with OpenCV

After installing OpenCV in the previous chapter, it is now time to perform image-processing operations using the OpenCV library. In this chapter, we will cover the following topics:

- Image processing with OpenCV
- Viewing a video feed from the Pi camera
- Building an object-following robot

Technical requirements

There are no new technical requirements as such for this chapter, but you will require the following things to perform the examples:

- A red, green, or blue colored ball for detection
- A Pi camera and an ultrasonic sensor mounted on the robot

The code files for this chapter can be downloaded from `https://github.com/PacktPublishing/Hands-On-Robotics-Programming-with-Cpp/tree/master/Chapter07`.

Image processing with OpenCV

In this section, we will look at the important functions of the OpenCV library. After that, we will write a simple C++ program using the OpenCV libraries and perform different image-processing operations on an image.

Important functions in OpenCV

Before writing any OpenCV program, it is important to understand some of the main functions in OpenCV and the output that these functions can give us. Let's start by looking at the functions:

- `imread()`: The `imread()` function is used to read an image or a video feed from the Pi camera or webcam. Inside the `imread()` function, we have to provide the location of the image. If the image and program files are in the same folder, we only need to provide the name of the image. However, if the image is stored in a different folder, then we need to provide the complete path of the image inside the `imread` function. We store the image value from the `imread()` function inside a matrix (`Mat`) variable.

  If the image and `.cpp` files are in the same folder, the code will look as follows:

  ```
  Mat img = imread("abcd.jpg"); //abcd.jpg is the image name
  ```

  If the image and `.cpp` file are in different folders, the code will look as follows:

  ```
  Mat img = imread("/home/pi/abcd.jpg"); //abcd image is in
                                         // the Pi folder
  ```

- `imshow()`: The `imshow()` function is used to show or view the image:

  ```
  imshow("Apple Image", img);
  ```

  The `imshow()` function consists of two parameters, as follows:

  - The first parameter is the window text
  - The second parameter is the variable name of the image that is to be displayed

  The output of the `imshow()` function is as follows:

- `resize()`: The `resize()` function is used to resize the dimensions of the image. This function is generally used when users are working with multiple windows at the same time:

```
resize(img, rzimg, cvSize(400,400));   //new width is 400
                                       //and height is 400
```

This function consist of three parameters:

- The first parameter is the variable name of the original image (`img`) that is to be resized.
- The second parameter is the variable name of the new image (`rzimg`) that will be resized.
- The third parameter is `cvSize`, and in this we enter the **new width** and the **height value**.

The output of the `resize()` function is as follows:

- `flip()`: This function is used to flip the image either horizontally, or vertically, or both at the same time:

```
flip(img, flipimage, 1)
```

This function consists of three parameters:

- The first parameter (`img`) is the variable name of the original image.
- The second parameter (`flipimage`) is the variable name of the flipped image.
- The third parameter is the flipping type; `0` denotes a vertical flip, `1` represents a horizontal flip, and `-1` means that the image should flip both horizontally and vertically.

The output of the `flip()` function is as follows:

- `cvtColor()`: This function is used to convert a normal RGB-colored image into a gray-scale image:

    cvtColor(img, grayimage, COLOR_BGR2GRAY)

This function consists of three parameters:

- The first parameter (`img`) is the variable name of the original image
- The second parameter (`grayimage`) is the variable of the new image that will be converted to gray-scale
- The third parameter, `COLOR_BGR2GRAY`, is the conversion type; BGR is RGB written in reverse

The output of the `cvtColor()` function is as follows:

- `threshold()`: The thresholding method is used to separate out regions of an image that represent an object. In simple terms, thresholding is used to recognize a particular object in an image. The thresholding method takes a source image (`src`), the thresholding value, and the maximum threshold value (255) as an input. It produces an output image (`thresimg`) by comparing the pixel values of the source image to the threshold value:

```
threshold(src, thresimg, threshold value, max threshold value,
threshold type);
```

The threshold function consists of five parameters:

- The first parameter (`src`) is the variable name of the image that is to be thresholded.
- The second parameter (`thresimg`) is the variable name of the thresholded image.
- The third parameter (`threshold value`) is the threshold value (from 0 to 255).
- The fourth parameter (`max threshold value`) is the maximum threshold value (255).
- The fifth parameter (`threshold type`) is the thresholding type.

Generally, there are five types of thresholding, as follows:

- **0-binary**: Binary thresholding is the simplest form of thresholding. In this thresholding, if any pixel on the source image (`src`) has a value greater than the threshold value, then in the output image (`thresimg`), this pixel is set to the maximum threshold value (255), and it will turn white. On the other hand, if any pixel on the source image has a value less than the threshold value, then in the output image the pixel value is set to 0, and it will appear black.

  For example, in the following code, the thresholded value is set to 85, the maximum threshold value is 255, and the thresholding type is a binary represented by the number 0:

  ```
  threshold(src, thresimg, 85, 255, 0);
  ```

So, if any pixels on the **Apple Image** source image have a value greater than the threshold value (that is, greater than 85), then those pixels will turn white in the output image. Similarly, the pixels on the source image whose values are less than the threshold value will turn black in the output image:

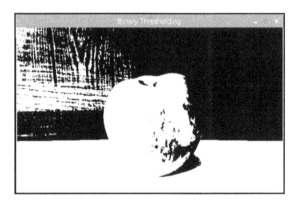

Binary thresholding

- **1-binary inverted**: Binary inverted thresholding is exactly the opposite of binary thresholding. In this type of thresholding, the pixel of the output image will turn black (0) if the source image's pixel value is greater than the threshold value, and will turn white (255) if the pixel value of the source image is less than the threshold value:

Binary inverted thresholding

- **2-truncated thresholding**: In truncated thresholding, if any pixel value on the `src` source image is greater than the threshold value, then in the output image, this pixel will be set to the threshold value. On the other hand, if any pixel value on the `src` source image is less than the threshold value, then in the output image, the pixel will retain its original color value:

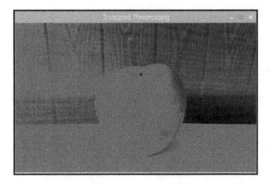

Truncated thresholding

- **3-threshold to zero**: In this thresholding, if any pixel value on the `src` source image is greater than the threshold value, then in the output image, the pixel will retain its original color value. On the other hand, if any pixel value on the `src` source image is less than the threshold value, then in the output image the pixel will be set to 0 (that is, black):

Threshold to zero

- **4-threshold to zero inverted**: In this thresholding, if any pixel value on the `src` is greater than the threshold value, then in the output image, the pixel will be set to `0`. If any pixel value on the `src` is less than the threshold value, then in the output image, the pixel will retain its original color value:

Threshold to zero inverted

- `inRange()`: The `inRange()` function is an advanced form of the thresholding function. Inside this function, we have to enter the minimum and maximum RGB color values of the object that we want to recognize. The `inRange()` function consists of four parameters:
    - The first parameter (`img`) is the variable name of the image that is to be thresholded.
    - There are two `Scalar` functions. In the second parameter, which is the first `Scalar` function, we have to enter the minimum RGB color of the object.
    - In the third parameter, which is the second `Scalar` function, we will enter the maximum RGB color value of the object.
    - The fourth parameter (`thresImage`) represents the output of the threshold image:

```
inRange(img, Scalar(min B,min G,min R), Scalar(max
B,max G,max R),thresImage)
```

Image moments—the concept of image moments is borrowed from **moments**, which is used in both mechanics and statistics to describe the spatial distribution of a set of points. In image processing or computer vision, an image moment is used to find the **centroid** of a shape, which is the average mean of all the points in a shape. In simple terms, image moments are used to find the center of any object once we have segmented it from the entire image. For example, in our case, we might want to find the center of the apple. The **image moments formula** for calculating the center of an object from an image is as follows:

$$x = \frac{M10}{M00}, y = \frac{M01}{M00}$$

- x represents the width of image
- y represents the height of image
- $M10$ represents the sum of all the x values in the image
- $M01$ represents the sum of all the y values in the image
- $M00$ represents the entire area of the image

- `circle`: As the name suggests, this function is used to draw a circle. It takes five parameters as the input:
    - The first parameter (`img`) is the variable name of the image on which you have to draw the circle.
    - The second parameter (`point`) is the center (the x, y position) point for the circle.
    - The third parameter (`radius`) is the radius of the circle.
    - The fourth parameter (`Scalar(B,G,R)`) is for coloring the circle; we do this using the `Scalar()` function.
    - The fifth parameter (`thickness`) is the thickness of the circle:

```
circle(img, point, radius, Scalar(B,G,R),thickness);
```

Object recognition using OpenCV

Now that we've understood the important functions of OpenCV, let's write a program to detect a colored ball from an image. Before we start, the first thing we have to do is to take a proper photo of the ball. You can use any ball for this project, but just make sure that the ball has a single color (red, green, or blue color ball is highly recommended) and that it is not multicolored. I'm using a green ball for this project:

Capturing the image

To capture the image of your ball, place it on some dark surface. I have placed my green ball on a mobile phone case, which is black:

If your ball is black, or of a dark color, you can place the ball on a surface that has a light color. This is to make sure there is a very high contrast between the color of the ball and the color of the background, which will help us in thresholding later on.

While capturing the image, make sure that there are no white colored patches on the ball, as this may cause problems later on while thresholding:

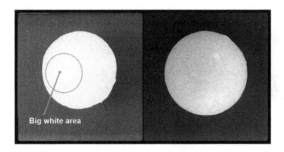

The photo on the left has a big white region because of too much lighting. On the right, the ball is lit up properly.

Once you are satisfied with the captured image, transfer it to your laptop.

Finding the RGB pixel values

We will now find the RGB pixel values of the ball by examining different points on it using the following steps:

1. Open Paint and open the saved image of the ball, as follows:

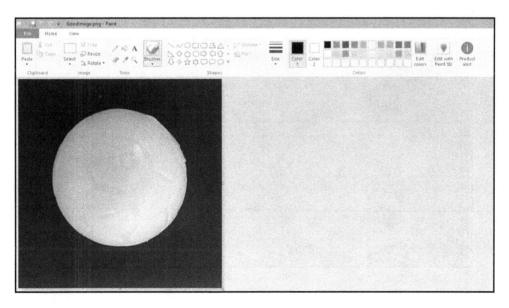

2. Next, using the color picker tool, take samples of the color by clicking anywhere on the ball:

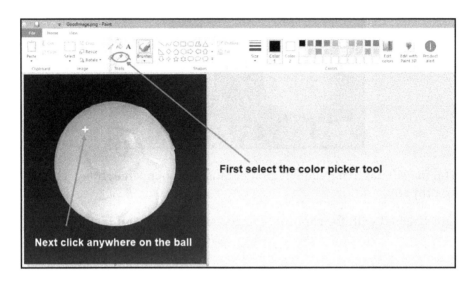

The **Color 1** box will show the sample of the color that was clicked on. In my case, this is green:

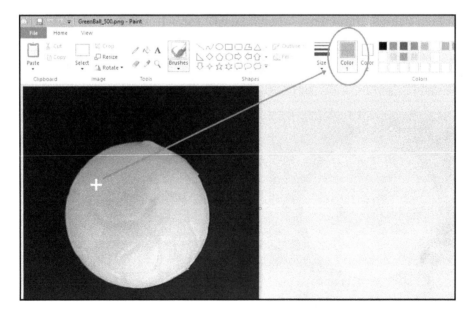

3. If you click on the **Edit Colors** option, you will see the RGB color value of that particular pixel. In my case, the RGB color value for the green pixel is **Red: 61**, **Green: 177**, and **Blue: 66**. Make a note of these values and save them for later:

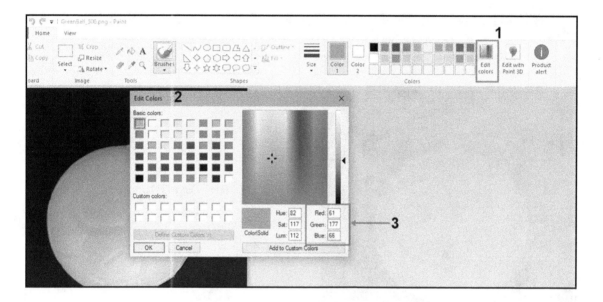

4. Now, select the color picker option again and click on another colored area of the ball to find out the RGB color value of that particular pixel. Again, note down this value. Do the same thing 13 to 14 times, making sure you include the lightest and the darkest colors on the ball:

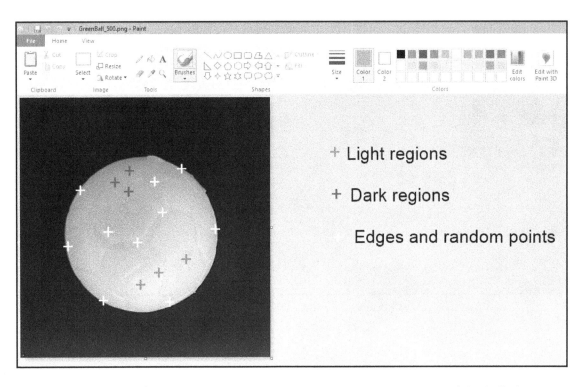

I have noted down the RGB values for six points on the edges of the ball, four points at random places around the ball, and six points on places where the color is either light green or dark green. After finding out the RGB values, highlight the lowest red, green, and blue color values and the highest red, green, and blue color values. We will use these values in our program later on to threshold the image.

5. You now need to transfer this image to your RPi. I transferred my image via Google Drive. I did this by uploading the image to Google Drive, next, opening the default Chromium web browser inside my RPi, signing into my Gmail account, opening Google Drive, and downloading the image.

The object detection program

The program for detecting the green ball is named `ObjectDetection.cpp` and I have saved it inside the `OpenCV_codes` folder. I have also copied the `greenball.png` image to this folder. You can download the `ObjectDetection.cpp` program from the `Chapter07` folder of the GitHub repository. So, the program for detecting the green ball is as follows:

```
#include <iostream>
#include<opencv2/opencv.hpp>
#include<opencv2/core/core.hpp>
#include<opencv2/highgui/highgui.hpp>
#include<opencv2/imgproc/imgproc.hpp>

using namespace cv;
using namespace std;

int main()
{

 Mat img, resizeimg,thresimage;
 img = imread("greenball.png");
 imshow("Green Ball Image", img);
 waitKey(0);

 resize(img, resizeimg, cvSize(640, 480));
 imshow("Resized Image", resizeimg);
 waitKey(0);

 inRange(resizeimg, Scalar(39, 140, 34), Scalar(122, 245, 119),
thresimage);
 imshow("Thresholded Image", thresimage);
 waitKey(0);

 Moments m = moments(thresimage,true);
 int x,y;
 x = m.m10/m.m00;
 y = m.m01/m.m00;
 Point p(x,y);
 circle(img, p, 5, Scalar(0,0,200), -1);
 imshow("Image with center",img);
 waitKey(0);

 return 0;
}
```

In the preceding program, we are importing four OpenCV libraries, which are `opencv.hpp`, `core.hpp`, `highgui.hpp`, and `imgproc.hpp`. We then declared the `cv` namespace that is a part of the OpenCV library.

The following is the explanation of the preceding program:

1. Inside the `main` function, we declared three matrix variables called `img`, `resizeimg`, and `thresimage`.
2. Next, the `greenball.png` file is read by the `imread()` function and stored in the `img` variable.
3. The `imshow("Green Ball Image", img)` line will display the image in a new window, as shown in the following screenshot:

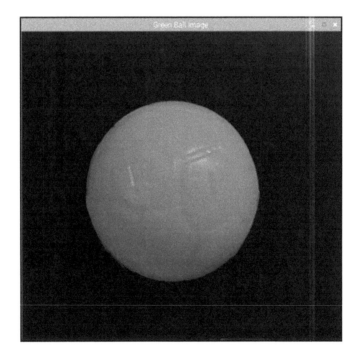

4. After this, the `waitKey(0)` function will wait for the keyboard input. It will then execute the next set of code. Once you press any key, the next two lines of code for resizing the image will execute.

5. The `resize` function will resize the width and height of the image so that the new width of the image is `640` and the height is `480`:

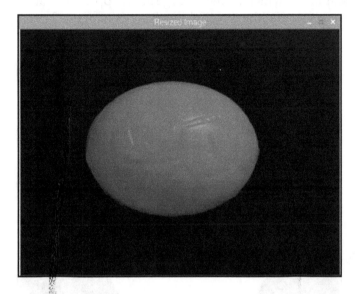

6. The thresholding operation is then performed using the `inRange` function. Inside the first `Scalar` function, I have entered the minimum RGB values for the green color of my ball, and in the second `Scalar` function, I have entered the maximum RGB values thereof. The thresholded image is stored in the `thresimage` variable.

 Inside the `Scalar` function, we have to enter the blue value first, followed by green, and then red.

7. After thresholding, the color of the ball will become white and the remaining portion of the image will become black. Some portions in the middle of the ball will appear black, which is fine. If large areas appear as black inside the white color, this means that the thresholding has not happened properly. In this case, you can try modifying the RGB values inside the `Scalar` function:

8. Next, using moments, we find the center of the object.

9. In the `moments(thresimage,true)` line, we have provided the `thresimage` variable as an input.

10. In the next three lines of code, we find the center of the white area and store that value in the point variable, `p`.

11. After that, to display the center of the ball, we use the `circle` function. Inside the circle function, we use the `img` variable as we will display the circular dot on the original image. Next, the point variable, `p`, tells the function where we have to display the dot. The width of the circular dot is set to `5`, and the color of the circular dot will be red, as we have only filled the last parameter of the `Scalar` function, which denotes the color red. If you want to set another color, you can change the color values inside the `Scalar` function:

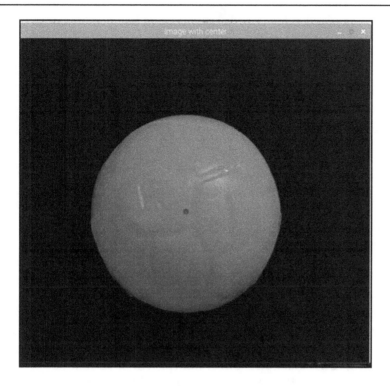

12. Press any key one more time and the final `waitKey(0)` function will close all of the windows apart from the Terminal window. To close the Terminal window, press *Enter*.

With the preceding program, we have learned how to resize, threshold, and generate a point (the red dot) on top of the image of the green ball. In the next section, we will perform some image recognition operations on a live video feed.

The OpenCV camera feed program

We will now write a simple C++ program to view the camera feed from the Pi camera. The program for viewing the video is as follows. The program is named `Camerafeed.cpp` and you can download it from the `Chaper07` folder of the GitHub repository:

```
int main()
{
 Mat videoframe;

VideoCapture vid(0);
```

```
if (!vid.isOpened())
 {
cout<<"Error opening camera"<<endl;
 return -1;
 }
 for(;;)
 {
 vid.read(videoframe);
 imshow("Frame", videoframe);
 if (waitKey(1) > 0) break;
 }
 return 0;
 }
```

The OpenCV libraries and namespace declaration is similar to that of the previous program:

1. First, inside the `main` function, we are declaring a matrix variable called `videoframe`.

2. Next, the `VideoCapture` datatype is used to capture a video feed from the Pi camera. It has a variable called `vid(0)`. The 0 number inside the `vid(0)` variable represents the index number of the camera. Currently, since we have only one camera attached to the RPi, the Pi camera will have an index of 0. If you attach a USB camera to the Raspberry Pi, then the USB camera will have an index of 1. By changing the index number, you can switch between the Pi camera and the USB camera.

3. Next, we specify that the `!vid.isOpened()` condition should be called if the camera is not able to capture any video feed. In this case, an `"Error opening camera"` message will be printed in the Terminal.

4. After that, the `vid.read(videoframe)` command will read the camera feed.

5. Using the `imshow("Video output", videoframe)` line, we can now view the camera feed.

6. The `waitKey` command will wait for keyboard input. Once you press any key, it will exit the code.

This is how you can view a video feed using the Pi camera.

Building an object-following robot

After thresholding an image and viewing the video feed from the Pi camera, we will combine both of these programs to create our object-following robot program.

In this section, we will write two programs. In the first program, we place the ball in front of the camera and trace it by creating a dot (using moments) in the center of the ball. Next, we will move the ball **up**, **down**, **left**, and **right** and note the point values at different positions on the camera.

In the second program, we will use these point values as an input and make the robot follow the ball object.

Ball tracing using moments

Before following the ball, the robot should first be able to trace it using the Pi camera. Before writing the program, let's see how we are going to track the ball.

Programming logic

First, we will resize the camera resolution to 640 x 480, as follows:

After resizing the width and height, we divide the camera screen horizontally into three equal sections:

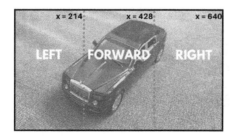

The **x coordinate values** from 0 to 214 represent the left section. The **x coordinate values** from 214 to 428 represent the forward section, while the **x coordinate values** from 428 to 640 represent the right section. We do not need to write any specific program to divide the camera screen into these three different sections, we just need to remember the minimum and maximum **x point values** for each of these sections.

Next, we will perform thresholding on the ball object. After this, we will use moments and generate a dot on the center of the ball. We will print the point value in the console and check the *x* and *y* point values at a particular segment of the screen:

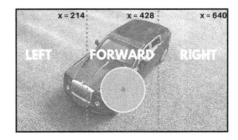

If the ball is in the **FORWARD** section, the **x coordinate value** must be between **214** and **428**. Since we are not dividing the screen vertically, we don't need to consider the *y* values. Let's now start with the ball tracing program.

The ball tracing program

The `BallTracing.cpp` program is as follows. You can download this program from the `Chapter07` folder of the GitHub repository:

```
int main()
{
  Mat videofeed, resizevideo, thresholdvideo;
  VideoCapture vid(0);
  if (!vid.isOpened())
  {
    return -1;
  }
  for (;;)
  {
    vid.read(videofeed);
  resize(videofeed, resizevideo, cvSize(640, 480));
  flip(resizevideo, resizevideo, 1);
  inRange(resizevideo, Scalar(39, 140, 34), Scalar(122, 245, 119),
thresholdvideo);
```

```
Moments m = moments(thresholdvideo,true);
int x,y;
x = m.m10/m.m00;
y = m.m01/m.m00;
Point p(x,y);
circle(resizevideo, p, 10, Scalar(0,0,128), -1);
imshow("Image with center",resizevideo);
  imshow("Thresolding Video",thresholdvideo);
cout<<Mat(p)<< endl;
if (waitKey(33) >= 0) break;
}
return 0;
}
```

Inside the main function, we have three matrix variables by the name of videofeed, resizevideo, and thresholdvideo. We have also declared a VideoCapture variable called vid(0) for capturing the video.

The following steps explain the BallTracing.cpp program in detail:

1. Inside the for loop, the vid.read(videofeed) code will read the camera feed.

2. Using the resize function, we resize the camera resolution to 640 x 480. The resized video feed is stored in the resizevideo variable.

3. Then, using the flip function, we flip the resized image horizontally. The output of the flipped video is again stored in the resizevideo variable. If we don't flip the video horizontally, the ball will appear as though it is moving on the right side when you move to the left and vice versa. If you have mounted the Pi camera upside down then you will need to flip the resized image vertically. To flip it vertically, set the third parameter to 0.

4. Next, with the inRange function, we threshold the video feed to make the colored ball stand out from the rest of the image. The thresholded video output is stored in the thresholdvideo variable.

5. Using moments, we find the center of the ball that is stored in the point variable, p.

6. Using the `circle` function, we display a red dot on the ball inside the
   `resizevideo` feed.

7. The first `imshow` function will display the resized (`resizedvideo`) video feed,
   while the second `imshow` function will display the thresholded
   (`thresholdvideo`) video feed:

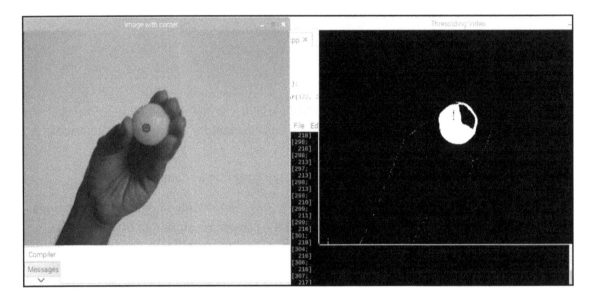

In the preceding screenshot, the left window shows the `resizevideo` feed and
we see the red dot on the green ball. The right window shows the threshold video
feed, in which only the region of the ball is appearing white.

8. Finally, the `cout<<Mat(p)<<endl;` code will display the *x* and *y* point values of
   the red dot inside the console. When you move the ball, the red dot will also
   move with it and the *x* and *y* position of the red dot will be displayed inside the
   console.

From the preceding screenshot, the values inside the square brackets, [298 ; 213], are
the point values. So the *x* value of the red dot in my case is in the range of 298 to 306 and
the *y* value is in the range of 216 to 218.

Setting up the object-following robot

After tracing the position of the ball, what remains is to make our robot follow the ball. We are going to use the *x* and *y* coordinate values as an input. While following the ball, however, we also have to make sure that the robot is at a suitable distance from the ball, so that it does not collide with the ball or the person who is holding it. To do this, we will also attach the ultrasonic sensor to our robot. For this project, I have attached the ultrasonic sensor **trigger** pin to the **wiringPi pin no 12**, and the **echo** pin to the **wiringPi pin no 13** via a **voltage divider circuit**.

Object-following robot program

The object-following robot program is basically a combination of the obstacle-avoiding program from Chapter 4, *Building an Obstacle-Avoiding Robot,* and the preceding ball-tracing program. The program is called ObjectFollowingRobot.cpp and you can download it from the Chapter07 folder of the GitHub repository:

```
int main()
  {
...
  float distance = (totalTime * 0.034)/2;

  if(distance < 15)
  {
  cout<<"Object close to Robot"<< " " << Mat(p)<< " " <<distance << " cm" <<
endl;
  stop();
  }

  else{
      if(x<20 && y< 20)
      {
      cout<<"Object not found"<< " " << Mat(p)<< " " <<distance << " cm" <<
endl;
      stop();
      }
      if(x > 20 && x < 170 && y > 20 )
      {
      cout<<"LEFT TURN"<< " " << Mat(p)<< " " <<distance << " cm" << endl;
      left();
      }
      if(x > 170 && x < 470)
      {
      cout<<"FORWARD"<< " " << Mat(p)<< " " <<distance << " cm" << endl;
      forward();
```

```
    }
    if(x > 470 && x < 640)
    {
    cout<<"RIGHT TURN"<< " " << Mat(p)<< " " <<distance << " cm" << endl;
    right();
    }


    }
    if (waitKey(33) >= 0) break;
    }
     return 0;
}
```

In the `main` function, after calculating the distance, thresholding the video, and placing the dot at the center of ball, let's take a look at the rest of the program:

1. The first `if` condition (`if(distance < 15)`), will check whether the robot is 15 cm away from the object. If the distance is less than 15 cm, the robot will stop. The forward, left, right, and stop functions are declared above the `main` function.

2. Underneath the `stop()` function, the `cout` statement will first print the message, `"Object close to Robot"`. After that, it will print the point(x,y) value (`Mat(p)`) and then the `distance` value. Inside every `if` condition, the `cout` statement will print the region (such as `LEFT`, `FORWARD`, or `RIGHT`), the point value, and the `distance` value.

3. If the distance is greater than 15 cm, the `else` condition will execute. Inside the `else` condition, there are three `if` conditions to find the position of the ball (by using the red dot on it as a reference).

4. Now, as soon as the camera is activated, or when the ball moves out of the camera's view, the red dot (the point) will reset to a position of **x:0, y:0**, which is at the extreme top-left of the screen. The first `if` condition (`if(x<20 && y< 20)`) inside the `else` block will check whether the position of the red dot is less than 20 on both the *x* and *y* axis. If it is, the robot will stop.

5. If the *x* position is between 20 and 170 and the *y* position is greater than 20, the red dot will be in the **LEFT** region and the robot will turn **LEFT**.

6. In this program, I have reduced the width of the **LEFT** and **RIGHT** regions and increased the width of the **FORWARD** region, as shown in the following photo. You can modify the width of each region according to your requirements:

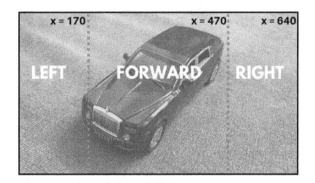

7. If the **x position** is between **170** and **470,** the red dot is in the **FORWARD** region and the robot will move **FORWARD**.

8. If the **x position** is between **470** and **640,** the red dot is in the **RIGHT** region and the robot will turn **RIGHT**.

Power up your robot using a power bank so that it can move freely. Next, compile the program and build it on your RPi robot. As long as the ball is not in front of the robot, the red dot will remain in the extreme top-left corner of the screen and the robot will not move. If you move the ball in front of the camera and if you are 15 cm away from the robot, the robot will start following the ball.

As the robot follows the ball, the color of the ball will vary because of external factors, such as sunlight or the light in the room. If the light in the room is low, the ball will appear a little darker to the robot. Similarly, if there is too much light in the room, some parts of the ball may also appear white. This may cause the thresholding to not work properly, which might mean the robot does not follow the ball smoothly. In this case, you will need to adjust the RGB values.

Summary

In this chapter, we looked at some of the important functions inside OpenCV libraries. After that, we put these functions to the test and recognized an object from an image. Next, we learned how to read a video feed from the Pi camera, how to threshold a colored ball, and how to place a red dot on top of it. Finally, we used the Pi camera and the ultrasonic sensor to detect the ball and follow it.

In the next chapter, we are going to expand our OpenCV knowledge by detecting human faces using the Haar Cascade. After that, we will recognize a smile and make the robot follow the face.

Questions

1. What is the process of separating an object from an image called?

2. What is the command to flip the image vertically?

3. If x>428 and y>320, what block will the red dot be in?

4. What is the command used for resizing the camera resolution?

5. If the object is not in front of the camera then where will the red dot be placed?

8
Face Detection and Tracking Using the Haar Classifier

In the previous chapter, we programmed the robot to detect a ball object and follow it. In this chapter, we will take our detection skills to the next level by detecting and tracking a human face, detecting human eyes, and recognizing a smile.

In this chapter, you will learn about the following topics:

- Face detection using the Haar cascade
- Detecting the eyes and smile
- Face-tracking robot

Technical requirements

In this chapter, you will need the following:

- Three LEDs
- A **Raspberry Pi (RPi)** robot (with the Raspberry Pi camera module connected to the RPi)

The code files for this chapter can be downloaded from `https://github.com/ PacktPublishing/Hands-On-Robotics-Programming-with-Cpp/tree/master/Chapter08`.

Face detection using the Haar cascade

Paul Viola and Micheal Jones proposed the Haar feature-based cascade classifier in their paper, *Rapid Object Detection using a Boosted Cascade of Simple Features*, in 2001. The Haar feature-based cascade classifier is trained using facial images as well as non-facial images. The Haar cascade classifier can not only detect a frontal face but can also detect the eyes, mouth, and nose of a person. The Haar feature-based classifier is also referred to as the Viola-Jones algorithm.

Basic working of the Viola-Jones algorithm

So, put simply, the Viola-Jones algorithm used Haar features to detect a face. Haar generally consists of two main features: **edge features** and **line features**. We will first understand these two features and then we will see how these features are used to detect faces:

- **Edge features**: This is generally used to detect edges. The edge feature consists of white and black pixels. Edge features can be further categorized into horizontal edge features and vertical edge features. In the following diagram, we can see the vertical edge feature on the left block and the horizontal edge feature on the right block:

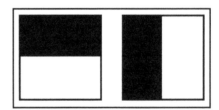

- **Line features**: This is generally used to detect lines. In line features, a white pixel is sandwiched between two black pixels, or a black pixel will be sandwiched between two white pixels. In the following diagram, you can see the two horizontal line features on the left, one below the other, and the vertical line features on the right, next to each other:

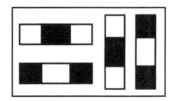

Face detection is always performed on a grayscale image, but this means that in a grayscale image, we may not have completely black and white pixels. So, let's refer to white pixels as brighter pixels and black pixels as darker pixels. If we look at the following grayscale face picture, the forehead region is lighter (brighter pixels) compared to the eyebrow region (darker pixels):

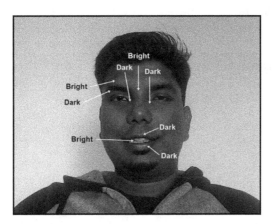

The area of the nose line is brighter compared to the eye and cheeks regions. Similarly, if we look at the mouth region, the upper lip region is darker, the teeth region is brighter, and the lower lip region is dark again:

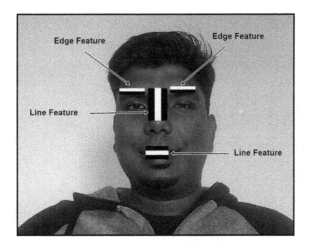

This is how, by using the edge and line features of the Haar cascade, we can detect the most relevant feature points in a human face, such as the eyes, nose, and mouth.

OpenCV 4.0 consists of different pre-trained Haar detectors, which can be used to detect a human face, including its eyes, nose, smile, and so on. Inside the Opencv-4.0.0 folder, there is a Data folder, and inside the Data folder, you will find the haarcascades folder. In this folder, you will find different Haar cascade classifiers. For frontal face detection, we will use the haarcascade_frontalface_alt2.xml detector. In the following screenshot, you can see the path of the haarcascades folders, with different Haar cascade classifiers present inside:

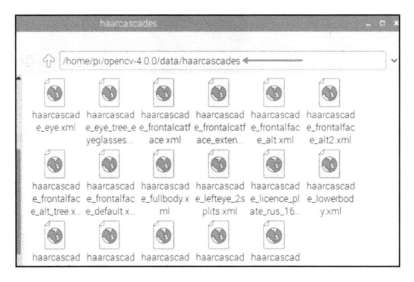

Now that we understand the basics of Viola-Jones features, we will program our robot to detect a human face using the Haar cascade.

Face-detection program

Let's write a program to detect a human face. I have named this program FaceDetection.cpp and you can download it from the Chapter08 folder of this book's GitHub repository.

Since we will be using haarcascade_frontalface_alt2.xml to detect faces, please make sure that the FaceDetection.cpp and haarcascade_frontalface_alt2.xml files are in the same folder.

To program face detection, follow these steps:

1. In the `FaceDetection.cpp` program, load the Haar's pre-trained frontal face XML using the `CascadeClassifier` class, as shown in the following code snippet:

```
CascadeClassifier
faceDetector("haarcascade_frontalface_alt2.xml");
```

2. Declare two matrix variables, called `videofeed` and `grayfeed`, along with a `VideoCapture` variable, called `vid(0)`, to capture footage from the RPi camera:

```
Mat videofeed, grayfeed;
VideoCapture vid(0);
```

3. Inside the `for` loop, read the camera feed. Then, flip the camera feed horizontally. Using the `cvtColor` function, we can convert our `videofeed` into grayscale. If your Pi camera is placed upside-down, set the third parameter inside `flip` function to 0. The `grayscale` output is stored in the `grayfeed` variable. The following code shows how to complete this step:

```
vid.read(videofeed);
flip(videofeed, videofeed, 1);
cvtColor(videofeed, grayfeed, COLOR_BGR2GRAY);
```

4. Let's perform a histogram equalization to improve the brightness and contrast of `videofeed`. Histogram equalization is required because sometimes, in low lighting, the camera may not be able to detect the face. To perform histogram equalization, we will use the `equalizeHist` function:

```
equalizeHist(grayfeed, grayfeed);
```

5. Let's detect some faces. For this, the `detectMultiScale` function is used, as follows:

```
detectMultiScale(image, object, scalefactor, min
neighbors, flags, min size, max size);
```

The `detectMultiScale` function that's shown in the preceding code snippet consists of the following seven parameters:

- `image`: Represents the input video feed. In our case, it is `grayfeed`, as we will detect the face from the grayscale video.
- `object`: Represents the vectors of a rectangle where each rectangle contains the detected faces.

- `scalefactor`: Specifies how much the image size must be reduced. The ideal value of scale factor is between 1.1 and 1.3.
- `flags`: This parameter can be set to CASCADE_SCALE_IMAGE, CASCADE_FIND_BIGGEST_OBJECT, CASCADE_DO_ROUGH_SEARCH, or CASCADE_DO_CANNY_PRUNING:
  - CASCADE_SCALE_IMAGE: This is the most popular flag; it informs the classifier that the Haar features for detecting the face are applied to the video or image
  - CASCADE_FIND_BIGGEST_OBJECT: This flag will tell the classifier to find the biggest face in the image or video
  - CASCADE_DO_ROUGH_SEARCH: This flag will stop the classifier once a face is detected
  - CASCADE_DO_CANNY_PRUNNING: This flag informs the classifier to not detect sharp edges, thus increasing the chances of face detection
- `min neighbors`: The minimum neighbors parameter affects the quality of the detected faces. Higher **min neighbor values** will recognize fewer faces, but whatever it detects will definitely be a face. Lower `min neighbors` values may recognize multiple faces, but sometimes it may also recognize objects that are not faces. The ideal `min neighbors` values for detecting faces is between 3 and 5.
- `min size`: The minimum size parameter will detect the minimum face size. For example, if we set the min size to 50 x 50 pixels, the classifier will only detect faces that are bigger than 50 x 50 pixels and ignore faces that are lower than 50 x 50 pixels. Ideally, we can set the min size to 30 x 30 pixels.
- `max size`: The maximum size parameter will detect the maximum face size. For example, if we set the max size to 80 x 80 pixels, the classifier will only detect faces that are smaller than 80 x 80 pixels. So, if you move too close to the camera and your face size exceeds the max size, your face will not be detected by the classifier.

6. Since the `detectMultiScale` function provides a vector of rectangles as its output, we have to declare a vector as the `Rect` type. The variable name as `face`. `scalefactor` is set to `1.1`, `min neighbors` is set to 5, and the minimum scale size is set 30 x 30 pixels. The max size is ignored here because if your face size becomes bigger than the max size, your face will not be detected. To complete this step, use the following code:

```
vector<Rect> face;
 faceDetector.detectMultiScale(grayfeed, faces, 1.3, 5, 0 |
CASCADE_SCALE_IMAGE, Size(30, 30));
```

After detecting faces, we will create a rectangle around the detected faces and display text on the top-left side of the rectangle that states "Face detected":

```
for (size_t f = 0; f < face.size(); f++)
 {
rectangle(videofeed, face[f], Scalar(255, 0, 0), 2);
putText(videofeed, "Face Detected", Point(face[f].x, face[f].y),
FONT_HERSHEY_PLAIN, 1.0, Scalar(0, 255, 0), 2.0);
 }
```

Inside the `for` loop, we are determining how many faces are detected using the `face.size()` function. If one is detected, `face.size()` equals 1, and the `for` loop will be satisfied. Inside the `for` loop, we have the rectangle and `putText` function.

The rectangle function will create a rectangle around the detected face. It consists of four parameters:

- The first parameter represents the image or video feed on which we want to draw the rectangle, which in our case is `videofeed`
- The second parameter of `face[f]` represents the detected face on which we have to draw the rectangle
- The third parameter represents the color of the rectangle (for this example, we have set the color to blue)
- The fourth and final parameter represents the thickness of the rectangle

The `putText` function is used to display text in an image or video feed. It consists of seven parameters:

- The first parameter represents the image or video feed on which we want to draw the rectangle.
- The second parameter represents the text message that we want to display.

- The third parameter represents the point on which we want the text to be displayed. The `face[f].x` and `face[f].y` functions represent the top-left point of the rectangle, so the text will be displayed on the top-left side of the rectangle.
- The fourth parameter represents the font type, which we have set to `FONT_HERSHEY_PLAIN`.
- The fifth parameter represents the font size of the text, which we have set to `1`.
- The sixth parameter represents the color of the text, which is set to green (`Scalar(0,255,0)`).
- The seventh and final parameter represents the thickness of the font, which is set to `1.0`.

Finally, using the `imshow` function, we will view the video feed, along with the rectangle and text:

```
imshow("Face Detection", videofeed);
```

After using the preceding code, and if you have compiled and built the program, you will see that a rectangle has been drawn around the detected face:

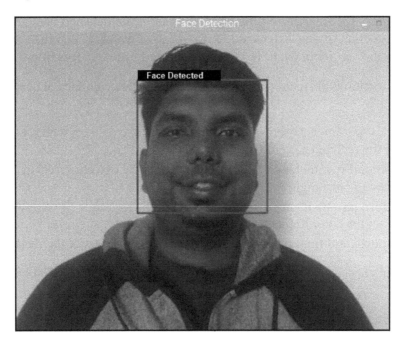

Next, we will detect human eyes as well as recognize a smile. Once the eyes and smile have been recognized, we will create circles around them.

Detecting the eyes and smile

The program for detecting the eyes and smile is called `SmilingFace.cpp`, and you can download it from the `Chapter08` folder of this book's GitHub repository.

Detecting the eyes

The `SmilingFace.cpp` program is basically an extension of the `FaceDetection.cpp` program, meaning that we will first find the region of interest, which is the face. Next, using the Haar `CascadeClassifier` for the eyes, we will detect the eyes and then draw circles around them.

Before writing the program, let's first understand the different eye `CascadeClassifier` that are available. OpenCV 4.0 has three main eye cascade classifiers:

- `haarcascade_eye.xml`: This classifier will detect both of the eyes simultaneously
- `haarcascade_lefteye_2splits.xml`: This classifier will detect only the left eye
- `haarcascade_righteye_2splits.xml`: This classifier will detect only the right eye

Depending on your requirements, you can use the `haarcascade_eye` classifier to detect both of the eyes, or you can use the `haarcascade_lefteye_2splits` classifier to detect only the left eye and the `haarcascade_righteye_2splits` classifier to detect only the right eye. In the `SmilingFace.cpp` program, we will first test the output with the `haarcascade_eye` classifier and then we will test the output with the `haarcascade_lefteye_2splits` and `haarcascade_righteye_2splits` classifiers.

Eye detection using haarcascade_eye

To test the `haarcascade_eye` output, observe the following steps:

1. Load this classifier inside our program:

   ```
   CascadeClassifier eyeDetector("haarcascade_eye.xml");
   ```

2. To detect the eyes, we need to find the face region (region of interest) in the image (video feed). Inside the face-detection `for` loop, we will create a `Mat` variable called `faceroi`. `videofeed(face[f])`, which will find faces in `videofeed` and store them inside the `faceroi` variable:

   ```
   Mat faceroi = videofeed(face[f]);
   ```

3. Create a vector of the `Rect` type, called `eyes`, and then use the `detectMultiScale` function to detect the eye region:

   ```
   vector<Rect> eyes;
   eyeDetector.detectMultiScale(faceroi, eyes, 1.3, 5, 0
   |CASCADE_SCALE_IMAGE,Size(30, 30));
   ```

   In the `detectMultiScale` function, the first parameter is set to `faceroi`, which means that we want to detect the eyes from the face region only and not from the entire video feed. The detected eyes will be stored in the eyes variable.

4. To create circles around the eyes, we will use a `for` loop. Let's find the center of the eyes. To find the center of an eye, we will use the `Point` datatype, and the equation inside the `eyecenter` variable will give us the center of the eye:

   ```
   for (size_t e = 0; e < eyes.size(); e++)
   {
   Point eyecenter(face[f].x + eyes[e].x + eyes[e].width/2,
   face[f].y + eyes[e].y + eyes[e].height/2);
   int radius = cvRound((eyes[e].width + eyes[e].height)*0.20);
   circle(videofeed, eyecenter, radius, Scalar(0, 0, 255), 2);
   }
   ```

The results of this can be seen here:

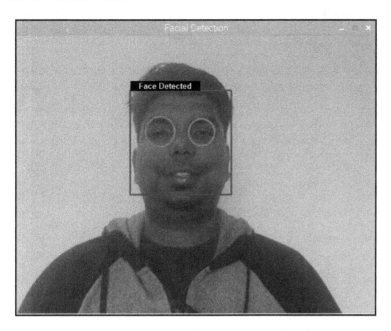

Using the `radius` variable, we have calculated the radius of the circle and then used the `circle` function to create red circles around the eyes.

Eye detection using haarcascade_lefteye_2splits and haarcascade_righteye_2splits

After detecting both eyes using the `haarcascade_eye` classifier, let's try to detect only the left eye or the right eye by using the `haarcascade_lefteye_2splits` and `haarcascade_righteye_2splits` classifiers, respectively.

Detecting the left eye

To detect the left eye, perform these steps:

1. Load the `haarcascade_lefteye_2splits` cascade classifier inside our program:

```
CascadeClassifier
eyeDetectorleft("haarcascade_lefteye_2splits.xml");
```

2. Since we want to detect the left eye in the face region, we will create a `Mat` variable called `faceroi` and inside it we will store the face region value:

```
Mat faceroi = videofeed(face[f]);
```

3. Use the `detectMultiScale` function to create a vector of the `Rect` type called `lefteye` to detect the left eye region. The `min neighbors` parameter is set to `25` so that the classifier detects only the left eye. If we set `min neighbors` lower than 25, the `haarcascade_lefteye_2splits` classifier may also detect the right eye, which is not what we want. To complete this step, use the following code:

```
vector<Rect> lefteye;
eyeDetectorleft.detectMultiScale(faceROI, lefteye, 1.3, 25, 0
|CASCADE_SCALE_IMAGE,Size(30, 30));
 for (size_t le = 0; le < lefteye.size(); le++)
 {
 Point center(face[f].x + lefteye[le].x +
lefteye[le].width*0.5, face[f].y + lefteye[le].y +
lefteye[le].height*0.5);
 int radius = cvRound((lefteye[le].width +
lefteye[le].height)*0.20);
 circle(videofeed, center, radius, Scalar(0, 0, 255), 2);
 }
```

The output of the preceding code is as follows:

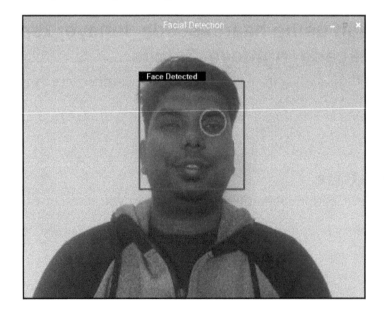

The `for` loop code for detecting the left and right eye separately is a part of the `SmilingFace.cpp` program, but it is commented. To test the code, first comment the `for` loop for detecting both the eyes simultaneously and then uncomment the other two `for` loops for detecting the left and right eyes.

Detecting the right eye

The programming logic for detecting the right eye is very similar to detecting the left eye. The only thing that we have to change is the classifier name and some variable names to distinguish the left and right eyes. To detect the right eye, perform these steps:

1. Load the `haarcascade_righteye_2splits` cascade classifier:

```
CascadeClassifier
eyeDetectorright("haarcascade_righteye_2splits.xml");
```

2. Inside the face-detection for loop, find the face region. Then, use the `detectMultiScale` function to detect the right eye. Use the `circle` function to create a green circle around the right eye. To do so, use the following code:

```
Mat faceroi = videofeed(face[f]);
vector<Rect>  righteye;
eyeDetectorright.detectMultiScale(faceROI, righteye, 1.3, 25, 0
|CASCADE_SCALE_IMAGE,Size(30, 30));
for (size_t re = 0; re < righteye.size(); re++)
 {
 Point center(face[f].x + righteye[re].x +
righteye[re].width*0.5, face[f].y + righteye[re].y +
righteye[re].height*0.5);
 int radius = cvRound((righteye[re].width +
righteye[re].height)*0.20);
 circle(videofeed, center, radius, Scalar(0, 255, 0), 2);
 }
```

The output of the preceding code is as follows:

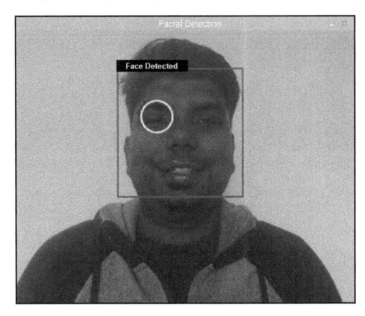

If we combine the left- and the right-eye detector code, the final output will be as follows:

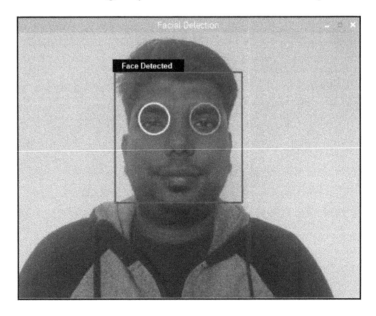

As we can see, the left eye in the picture is surrounded by a red circle and the right eye is surrounded by a green circle.

Recognizing a smile

After detecting the eyes from the face region, let's write the program to recognize a smiling face. The webcam will recognize a smiling face when it detects a black-white-black line feature around the mouth, that is, the upper and lower lip are generally a bit darker compared to the teeth region:

Programming logic for smile recognition

The programming logic for smile recognition is similar to eye detection, and we will also write the smile-recognition program inside the face-detection `for` loop. To program smile recognition, follow these steps:

1. Load the smile `CascadeClassifier`:

   ```
   CascadeClassifier smileDetector("haarcascade_smile.xml");
   ```

2. We need to detect the mouth region, which is inside the face region. The face region is again our region of interest, and to find the face region from the video feed, use will use the following command:

   ```
   Mat faceroi = videofeed(face[f]);
   ```

3. Declare a `smile` variable, which is a vector of the `Rect` type. Then, use the `detectMultiScale` function. In the `detectMultiScale` function, min neighbors is set to 25 so that a circle is created only when a person smiles (if we set min neighbor to lower than 25, a circle may be created around the mouth, even if the person is not smiling). You can vary the min neighbors value between 25-35. Next, inside the `for` loop, we have written the program to create a green circle around the mouth. To complete this step, use the following code:

   ```
   vector<Rect> smile;
   smileDetector.detectMultiScale(faceroi, smile, 1.3, 25, 0
   |CASCADE_SCALE_IMAGE,Size(30, 30));
   ```

```
for (size_t sm = 0; sm <smile.size(); sm++)
{
Point scenter(face[f].x + smile[sm].x + smile[sm].width*0.5,
face[f].y + smile[sm].y + smile[sm].height*0.5);
int sradius = cvRound((smile[sm].width +
smile[sm].height)*0.20);
circle(videofeed, scenter, sradius, Scalar(0, 255, 0), 2);
}
```

The output of the preceding code is as follows:

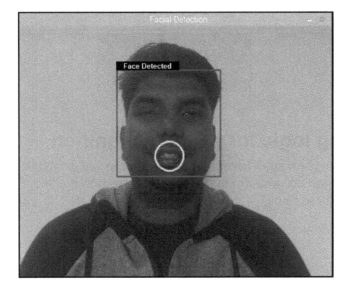

In the next section, we will turn on different LEDs on the robot when the eyes and smile are detected. We will also make our robot follow the detected face when the face moves.

Face-tracking robot

The program for turning LEDs on/off and tracking a human face is called Facetrackingrobot.cpp, and you can download it from the Chapter08 folder of this book's GitHub repository.

In the Facetrackingrobot program, we will first detect the face, and then the left eye, right eye, and smile. Once the eyes and smile are detected, we will turn the LEDs on/off. After this, we will create a small dot in the center of the face rectangle, and then use this dot as a reference to move our robot.

Wiring connections

For the `Facetrackingrobot` program, we will need a minimum of three LEDs: one for the left eye, one for the right eye, and one LED for smile recognition. The three LEDs are shown in the following diagram:

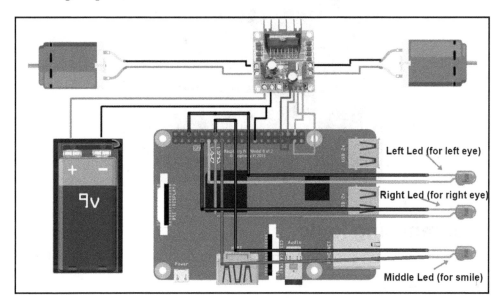

The wiring connections of the LEDs and the robot are as follows:

- The left LED, which corresponds to the **left eye**, is connected to **wiringPi pin 0**
- The right LED, which corresponds to the **right eye**, is connected to **wiringPi pin 2**
- The middle LED, which corresponds to a **smile**, is connected to **wiringPi pin 3**
- The **IN1** pin of the motor driver is connected to **wiringPi pin 24**
- The **IN2** pin of the motor driver is connected to **wiringPi pin 27**
- The **IN3** pin of the motor driver is connected to **wiringPi pin 25**
- The **IN4** pin of the motor driver is connected to **wiringPi pin 28**

On my robot, I have taped the left and right LEDs on the top chassis of the robot. The third LED (the middle LED) is taped to the bottom chassis of the robot. I have used green LEDs for the eyes and a red LED for the smile:

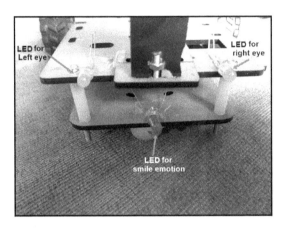

The programming logic

In the `Facetrackingrobot` program, wiringPi pins 0, 2, and 3 are set as output pins:

```
pinMode(0,OUTPUT);
pinMode(2,OUTPUT);
pinMode(3,OUTPUT);
```

From the face-detection program, you may have noticed that the face-tracking process is very slow. Therefore, when you move your face to the left or right, you have to make sure that the motors don't move too fast. To reduce the speed of the motor, we will use the `softPwm.h` library, which we also used in `Chapter 2`, *Implementing Blink with wiringPi*:

1. From the `softPwm.h` library, use the `softPwmCreate` function to declare the four motor pins (`24`, `27`, `25`, and `28`):

```
softPwmCreate(24,0,100); //pin 24 is left Motor pin
softPwmCreate(27,0,100); //pin 27 is left motor pin
softPwmCreate(25,0,100); //pin 25 is right motor pin
softPwmCreate(28,0,100); //pin 28 is right motor pin
```

The first parameter inside the `softPwmCreate` function denotes the wiringPi pin of the RPi. The second parameter represents the minimum speed at which we can move the motor, and the third parameter represents the maximum speed at which we can move the motor.

2. Load the face, left eye, right eye, and smile `CascadeClassifiers`:

```
CascadeClassifier
faceDetector("haarcascade_frontalface_alt2.xml");
CascadeClassifier
eyeDetectorright("haarcascade_righteye_2splits.xml");
CascadeClassifier
eyeDetectorleft("haarcascade_lefteye_2splits.xml");
CascadeClassifier smileDetector("haarcascade_smile.xml");
```

3. Inside the `for` loop, declare three Boolean variables, called `lefteyedetect`, `righteyedetect`, and `isSmiling`. Set all three variables to `false`. Using these three variables, we will detect whether the left eye, right eye, and smile are detected. Declare the `facex` and `facey` variables, which will be used to find the center of the face rectangle. To complete this step, use the following code:

```
bool lefteyedetect = false;
bool righteyedetect = false;
bool isSmiling = false;
int facex, facey;
```

4. Use the `detectMultiScale` function to detect the face and then, inside the `for` loop, we will write the program to create a rectangle around the detected face:

```
vector<Rect> face;
faceDetector.detectMultiScale(grayfeed, face, 1.1, 5, 0 |
CASCADE_SCALE_IMAGE,Size(30, 30));
 for (size_t f = 0; f < face.size(); f++)
 {
 rectangle(videofeed, face[f], Scalar(255, 0, 0), 2);

 putText(videofeed, "Face Detected", Point(face[f].x,
face[f].y), FONT_HERSHEY_PLAIN, 1.0, Scalar(0, 255, 0), 1.0);

 facex = face[f].x +face[f].width/2;
 facey = face[f].y + face[f].height/2;

 Point facecenter(facex, facey);
 circle(videofeed,facecenter,5,Scalar(255,255,255),-1);
```

`face[f].x + face[f].width/2` will return the *x* center value of the rectangle and `face[f].y + face[f].height/2` will return the *y* center value of the rectangle. The *x* center value is stored in the `facex` variable and the *y* center value is stored in the `facey` variable.

5. To find the center of the rectangle, provide `facex` and `facey` as input to the `Point` variable, called `facecenter`. Inside the circle function, use the `facecenter` point variable as an input to create a dot in the center of the face rectangle:

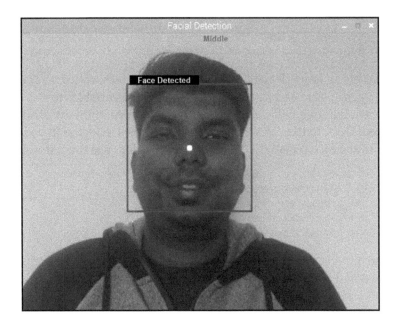

6. When the left eye is detected, we will create a red circle around it and set the `lefteyedetect` variable to `true`:

```
eyeDetectorleft.detectMultiScale(faceroi, lefteye, 1.3, 25, 0
|CASCADE_SCALE_IMAGE,Size(30, 30));
 for (size_t le = 0; le < lefteye.size(); le++)
 {
 Point center(face[f].x + lefteye[le].x +
lefteye[le].width*0.5, face[f].y + lefteye[le].y +
lefteye[le].height*0.5);
 int radius = cvRound((lefteye[le].width +
lefteye[le].height)*0.25);
 circle(videofeed, center, radius, Scalar(0, 0, 255), 2);
 lefteyedetect = true;
 }
```

7. When the right eye is detected, we will create a light blue circle around it and set the `righteyedetect` variable to `true`:

```
eyeDetectorright.detectMultiScale(faceroi, righteye, 1.3, 25,
0 |CASCADE_SCALE_IMAGE,Size(30, 30));
for (size_t re = 0; re < righteye.size(); re++)
{
Point center(face[f].x + righteye[re].x +
righteye[re].width*0.5, face[f].y + righteye[re].y +
righteye[re].height*0.5);
int radius = cvRound((righteye[re].width +
righteye[re].height)*0.25);
circle(videofeed, center, radius, Scalar(255, 255, 0), 2);
righteyedetect = true;
}
```

8. When the smile is detected, we will create a green circle around the mouth and set `isSmiling` to `true`:

```
smileDetector.detectMultiScale(faceroi, smile, 1.3, 25, 0
|CASCADE_SCALE_IMAGE,Size(30, 30));
for (size_t sm = 0; sm <smile.size(); sm++)
{
Point scenter(face[f].x + smile[sm].x + smile[sm].width*0.5,
face[f].y + smile[sm].y + smile[sm].height*0.5);
int sradius = cvRound((smile[sm].width +
smile[sm].height)*0.25);
circle(videofeed, scenter, sradius, Scalar(0, 255, 0), 2, 8,
0);
isSmiling = true;
}
```

In the following screenshot, you can see a red circle drawn around the left eye, a light blue circle drawn around the right eye, a green circle drawn around the mouth, and a white dot in the center of the blue rectangle that surrounds the face:

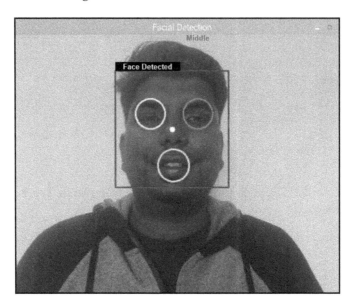

Using three `if` conditions, we will check when the `lefteyedetect`, `righteyedetect`, and `isSmiling` variables are `true`, and when they are `true`, we will turn on their respective LEDs:

- The `lefteyedetect` variable will be `true` when the left eye is detected. When the left eye is detected, we will turn on the left LED on the robot, which is connected to wiringPi pin 0, as shown in the following code:

```
if(lefteyedetect == true){
digitalWrite(0,HIGH);
}
else
{
digitalWrite(0,LOW);
}
```

- The `righteyedetect` variable will be `true` when the right eye is detected. When the right eye is detected, we will turn on the right LED on the robot, which is connected to wiringPi pin 2:

```
if(righteyedetect == true){
digitalWrite(2,HIGH);
}
else
{
digitalWrite(2,LOW);
}
```

- Finally, the `isSmiling` variable will be true when the smile is recognized. When the smile is recognized, we will turn on the middle LED, which is connected to wiringPi pin 3:

```
if(isSmiling == true){
 digitalWrite(3,HIGH);
 }
else
{
digitalWrite(3,LOW);
}
```

Next, we will use the white dot (point) on the face rectangle to move the robot left and right.

Using the white dot on the face triangle to move the robot

Similar to `Chapter 7`, *Building an Object-Following Robot with OpenCV*, we will divide the camera screen into three sections: the left section, the middle section, and the right section. When the white dot is in the left or right section, we will turn the robot left or right, thus tracking the face. Even though I have not resized `videofeed`, the resolution of `videofeed` is set to 640 x 480 (a width of 640 and a height of 480).

You can vary the range as per your requirements, but as shown in the following diagram, the left section is set to an x range from 0 to 280, the middle section is set to a range of 280-360, and the right section is set to a range of 360 to 640:

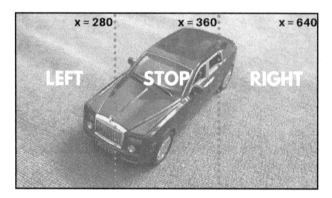

When we move our face, the face rectangle will move, and when the face rectangle moves, the white dot in the center of the rectangle will also move. When the dot moves, there will be changes in the `facex` and `facey` values. When dividing the camera screen into three sections, we will use the `facex` variable as a reference, and then we will use three if conditions to check which section the white dot is in. The code for comparing the `facex` values is as follows:

```
if(facex > 0 && facex < 280)
 {
 putText(videofeed, "Left", Point(320,10), FONT_HERSHEY_PLAIN, 1.0,
CV_RGB(0, 0, 255), 2.0);
 softPwmWrite(24, 0);
 softPwmWrite(27, 30);
 softPwmWrite(25, 30);
 softPwmWrite(28, 0);
 }

if(facex > 360 && facex < 640)
 {
 putText(videofeed, "Right", Point(320,10), FONT_HERSHEY_PLAIN, 1.0,
CV_RGB(0, 0, 255), 2.0);
 softPwmWrite(24, 30);
 softPwmWrite(27, 0);
 softPwmWrite(25, 0);
 softPwmWrite(28, 30);

 }
if(facex > 280 && facex < 360)
 {
```

```
putText(videofeed, "Middle", Point(320,10), FONT_HERSHEY_PLAIN, 1.0,
CV_RGB(0, 0, 255), 2.0);
 softPwmWrite(24, 0);
 softPwmWrite(27, 0);
 softPwmWrite(25, 0);
 softPwmWrite(28, 0);
 }
```

If the first `if` condition is satisfied, this means that the white dot is between 0 to 280. In this case, we are printing the `Left` text on `videofeed` and then using the `softPwmWrite` function so that the robot will take an axial left turn. Inside the `softPwmWrite` function, the first parameter represents the pin number and the second parameter represents the speed at which our motor will move. Since wiringPi pin 24 is set to 0 (low), and wiringPi pin 27 is set to 30, the left motor will move backwards with a speed of 30. Similarly, since wiringPi pin 25 is set to 30, and wiringPi pin 28 is set to 0 (low), the right motor will move forward with a speed of 30.

The speed value of 30 is in the range of 0 to 100, which we set in the `softPwmCreate` function. You can also vary the speed value.

If the white dot is between 360 and 640, the `Right` text will be printed and the robot will take an axial right turn at a speed of 30.

Finally, when the white dot is between 280 and 360, the `Middle` text will be printed and the robot will stop moving.

This is how we can make the robot track a face and follow it.

Summary

In this chapter, we used the Haar face classifier to detect a face from a video feed and then we drew a rectangle around it. Next, we detected the eyes and smile from a given face, and drew circles around the eyes and mouth. After this, using our knowledge of face, eye, and smile detection, we turned the LEDs of our robot on and off when the eyes and smile were detected. Finally, by creating a white dot in the center of the face rectangle, we made the robot follow our face.

In the next chapter, we will learn how to control our robot using our voice. We will also create an Android application that will recognize what we are speaking about. When the Android application detects certain keywords, the Android smartphone's Bluetooth will send bits of data to the Raspberry Pi Bluetooth. Once our robot recognizes these keywords, we will use them to move the robot in different directions.

Questions

1. What is the name of the classifier that we use to detect faces?
2. When we open the mouth, which type of feature is created?
3. Which cascade can be used to detect only the left eye?
4. When detecting the eyes from the face, what is that region generally referred to as?
5. What is the use of the `equalizeHist` function?

4
Section 4: Smartphone-Controlled Robot

In this section, we will use the AppInventor2 website and create an Android app, using which you can control the robot with your smartphone.

The following chapter is included in this section:

- Chapter 9, *Building a Voice-Controlled Robot*

Building a Voice-Controlled **9** Robot

In 2012, I wanted to create a robot that could be controlled by an Android smartphone. At that time, however, I did not know much about Android programming. To my surprise, I came across an amazing website called **App Inventor** (http://www.appinventor.org/), which allows users to develop Android applications by joining programming blocks in the same way as joining pieces in a puzzle.

In this final chapter, we will use the App Inventor website and learn to control our robot with an Android smartphone using our voice as an input. We will cover the following topics:

- An introduction to App Inventor
- Creating a voice application
- Pairing the Android smartphone and **Raspberry Pi (RPi)** via Bluetooth
- Developing the Bluetooth program for RPi

Technical requirements

- An Android smartphone running on Android version Lollipop (version number 5.0-5.1.1) or above
- Raspberry Pi robot

The code files for this chapter can be downloaded from `https://github.com/ PacktPublishing/Hands-On-Robotics-Programming-with-Cpp/tree/master/Chapter09`.

An introduction to App Inventor

App Inventor is an open source web-based application that was originally developed by Google. It is currently maintained by the **Massachusetts Institute of Technology** (**MIT**). It allows users to develop Android applications using its state-of-the-art graphical programming interface, which is similar to Scratch. Developers have to drag and drop visual blocks to create an Android app with App Inventor. The current version of App Inventor is referred to as **App Inventor 2** (version 2), or **AI2**.

In the following diagram you can see how each programming block is connected to each other like a puzzle:

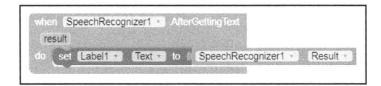

In this section, we'll look at how to create an App Inventor account and then create our first Android app using App Inventor.

Creating a Talking Pi Android app

Talking Pi is a simple Android application in which you type text inside a textbox and the smartphone displays and reads out the text. Before creating this Android app, we first need to get access to the App Inventor 2 dashboard. The final layout of the Talking Pi application will look somewhat like this:

To create Android applications using App Inventor 2, you must have a Gmail account. If you have one already, sign in on a browser of your choice. If you do not have one, you will need to create one. Let's now look at the steps for linking App Inventor 2 with your Gmail account:

1. After signing in, go to the following link: `ai2.appinventor.mit.edu/`. If you have signed in with multiple Gmail accounts inside your browser, you'll need to select one ID in particular:

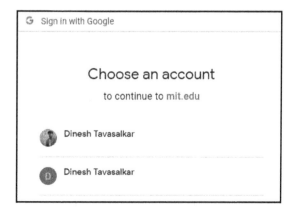

2. Next, you'll need to agree to the AI2 terms of service. You'll then be ready to create your Talking Pi application. To create a new Android application project, click on the **Start new project** button as follows:

3. Next, name the project `TalkingPi` and then click **OK:**

After creating your project, you will see the following four main panels inside App Inventor called **Palette**, **Viewer**, **Components**, and **Properties:**

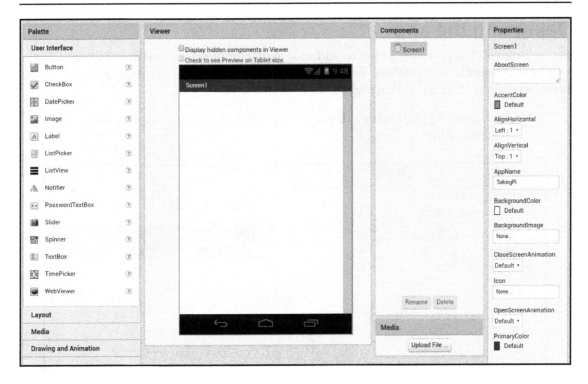

Let's now understand the workings of each of these panels as follows:

- The **Palette** panel consists of different components such as buttons, textboxes, canvases, Bluetooth, a video player, and so on.
- The **Viewer** panel consists of a screen in which we can drag and drop UI components from the **Palette**.
- The **Components** panel displays a list of visible and non-visible components that are added inside the screen. A button, for example, is a visible component, as it is visible on the screen. On the other hand, Bluetooth is a non-visible component as it is not visible on the screen, but it functions in the background. All invisible components are displayed below the screen.
- The **Properties** panel allows us to modify the properties of the components that are selected in the **Components** panel.

Let's now move on to designing the app.

Designing the app

In our Talking Pi app, we are going to add four main components: **TextBox**, **Button**, **Label**, and **TextToSpeech**. The **TextBox**, **Button**, and **Label** components are inside the **User Interface** option. Take the following steps:

1. You can drag the **TextBox**, **Button**, and **Label** components one by one inside the screen as follows:

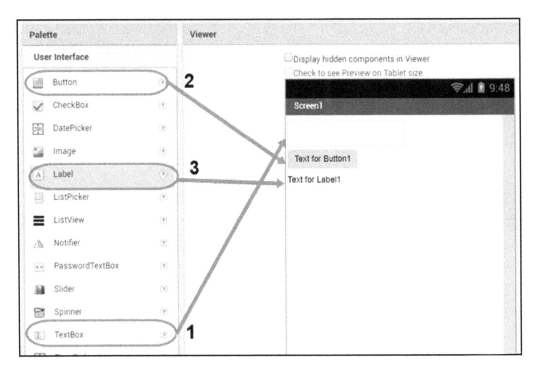

2. After adding these three components, you will notice that they are aligned to the top-left of the screen, which looks a bit odd. To position them horizontally in the center of the screen, select **Screen1** from the **Components** panel and change **AlignHorizontal** to **Center**, as shown in the following screenshot:

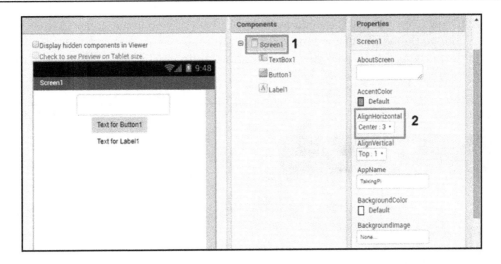

3. Next, to add some spacing between the three components, we can add layout components in between the **TextBox**, **Button**, and **Label** components. You can choose either a **HorizontalArrangement** or **VerticalArrangement**, for example:

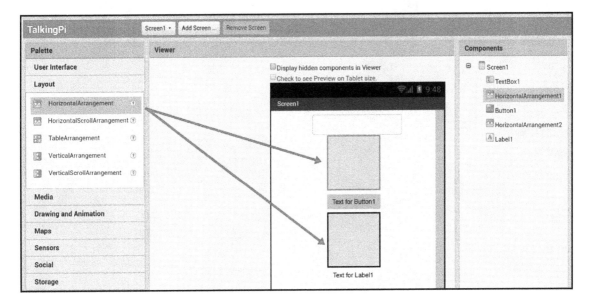

4. If you want to vary the distance between two components, you will need to change the height of the **HorizontalArrangement**. To do this, select the **Height** property and set a particular pixel value on the **HorizontalArrangement**, as follows:

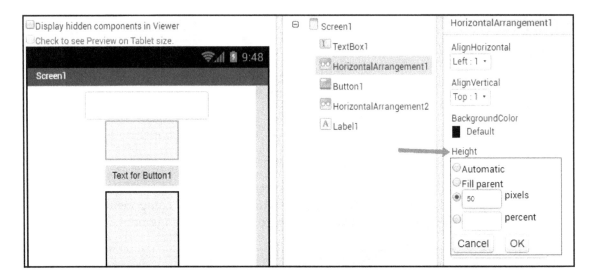

5. Next, select the **Button1** component and change its text to CLICK HERE:

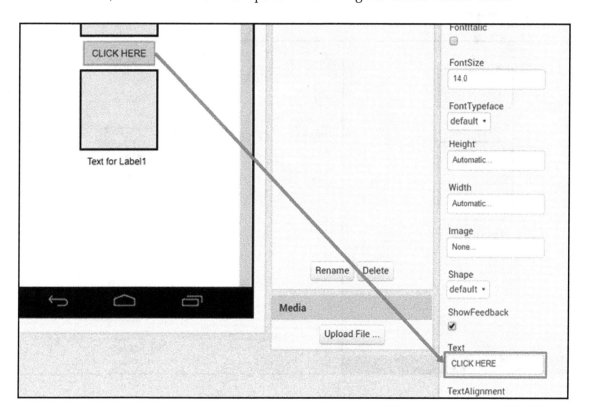

6. Similarly, select the **Label1** component, change its text to `TalkingPi`, and increase its **FontSize** to `24`:

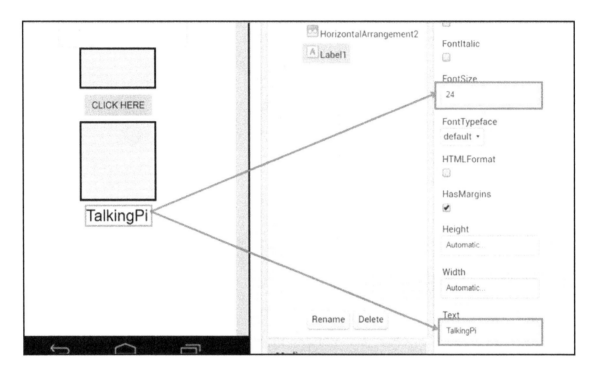

7. Finally, open the **Media** option and drag the **TextToSpeech** component on to the screen. Since the **TextToSpeech** component is a non-visible component, it will appear below the screen as follows:

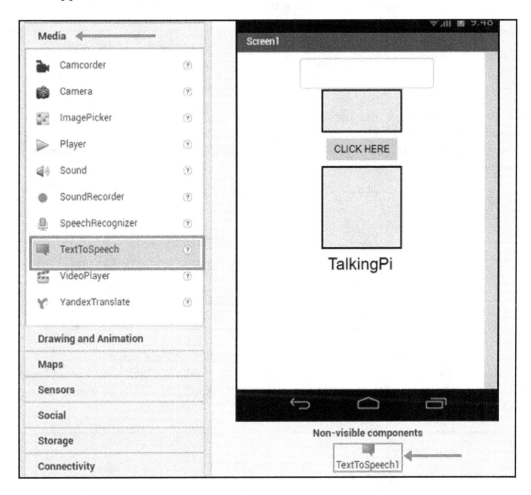

We have now basically finished designing the Talking Pi application. Let's now go inside the **Blocks** option and create the programming blocks for displaying the text and translating it into speech at the click of a button.

Programming the blocks

After designing the UI for the app, click on the **Blocks** button, which is next to the **Design** button, as shown in the following screenshot:

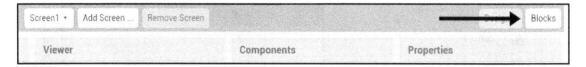

Inside the block section, on the left side you will see **Screen1**, which contains all the components (both visible and non-visible) that we have dragged onto the screen. If you click on any of the components, you will notice the following types of blocks for each component:

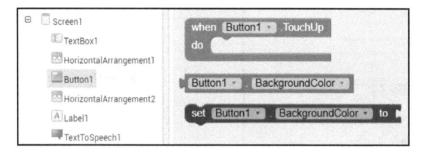

We will focus our attention mainly on the three types of block that make up each component. We will refer to these as **Main block**, **Intermediate block**, and **Final block**. Each of these blocks must be connected in the correct sequence to get a proper working program, as shown in the following diagram:

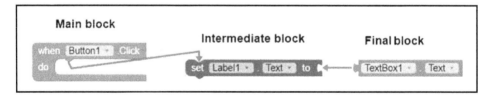

Let's take a look at each block.

Main block

The **main block** inside App Inventor 2 is similar to a **when** loop, which indicates an action to carry out when something happens. The **main block** is always connected to an **intermediate block**. We cannot connect the **final block** directly to the **main block**. The main block consists of a drop-down menu from which we can choose between multiple components that are of the same type. For example, look at the following screenshot:

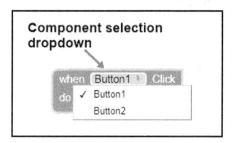

You can see that in a scenario where you have multiple buttons, you can choose a particular button from the drop-down list.

Intermediate block

The **intermediate block** consists of an **input socket** and an **output socket**. The **input socket** is connected to the **main block** and the **output socket** is connected to the **final block**, as shown in the following diagram:

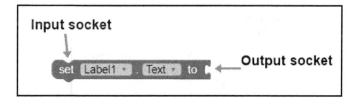

The **intermediate block** consists of two drop-down menus. The first drop-down menu represents the components of the same type. For example, if you have multiple labels, you can choose a particular label from the first drop-down menu. The second drop-down menu represents the properties of the component. For **Label1**, for example, we have **Text**, **BackgroundColor**, **Width**, and so on, as shown in the following diagram:

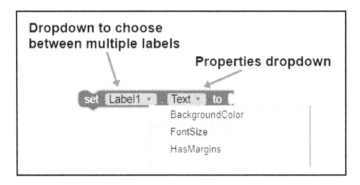

For example, **Label1** . **Text** means that we want to set or change the text of **Label1**.

Final block

The **final block** is connected to the **intermediate block**. It also consists of two drop-down menus from which we can select a particular component and its specific properties, as shown in the following screenshot:

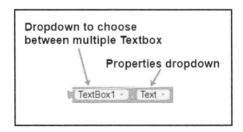

We will be using these three types of blocks to create our Talking Pi program. Let's start with the block programming.

The Talking Pi program

The Talking Pi programming logic is very simple. When **Button1** is pressed, **Label1** must display the text that is typed inside **Textbox1** and **TextToSpeech1** must read out that text. The steps for executing this block program are as follows:

1. First, click on the **Button1** component and select the **when Button1.Click** main block:

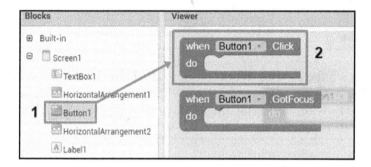

2. Next, since we want to change the text of **Label1** when **Button1** is clicked, choose the **Label1.Text** block from the **Label1** component:

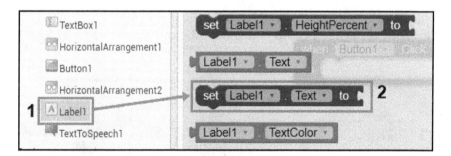

3. Next, drag the **Label1.Text** block inside the **Button1.Click** block to join both the blocks. Once both the blocks are joined, you will hear a click sound:

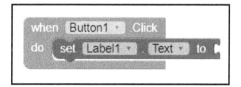

4. Now, we want to display the text inside the TextBox in the label component. From the **TextBox1** component, select the **TextBox1.Text** block as follows:

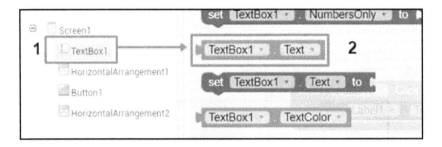

5. Next, attach the **TextBox1.Text** block to the **Label1.Text** block. Now, when you press the button, the label will display the text that is written inside the Textbox. **Label1** is now set to display the text that is inside **TextBox1**, as follows:

6. After this, to read out the text that is inside the textbox, click on the **TextToSpeech1** component and choose the **call TextToSpeech1.Speak** block, as follows:

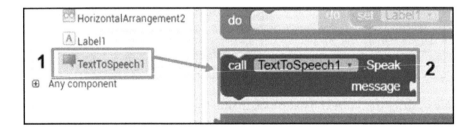

7. Connect this block below the **Label1.Text** block. Inside the message socket, connect the **TextBox1.Text** final block. This means that whatever text is written inside the textBox will be spoken as a message by the **TextToSpeech1** block, for example:

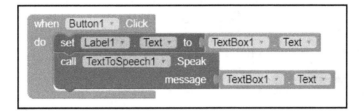

We have now finished designing our block program. To build and run this app inside your Android smartphone, click on the **Build** drop-down menu and choose between the two build types, as illustrated in the following diagram:

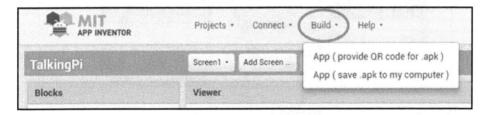

The first option, **App (provide QR Code for .apk)**, will generate a QR code that you can scan with your Android smartphone (using a QR scanner app). After scanning the QR code, the .apk file of the application will be downloaded inside your Android smartphone. Install the .apk file to test the output of the application.

The second option, **App (save .apk to my computer)**, will generate and download a .apk file inside your computer. You will need to transfer the .apk file from your computer to your smartphone and install the .apk file. I personally prefer the first option as the .apk file is directly downloaded inside the smartphone.

You can also download the MIT AI2 companion application from the Android play store and install it inside your Android smartphone to test the application in real time. So in the play store, search for **MIT AI2 Companion** and then click on the **Install** button to install the app. The app page is shown in the following screenshot:

After installing the MIT AI2 Companion app inside your Android smartphone, click on the **Scan QR code** button or enter the six-digit alphabetical code (next to the QR code) inside the MIT AI2 Companion app, and then click on the **Connect with code** button. To generate the QR code or the six-digit number, click on **Connect** and then select **AI Companion**, as follows:

Importing and exporting the .aia file of the app

You can export your Android application by generating its `.aia` file. To create a `.aia` file, follow any one of these steps:

- Click on **Projects** and then select **Export selected project (.aia) to my computer**:

- Similarly, you can click on **Projects** and then select **Import project (.aia) from my computer**:

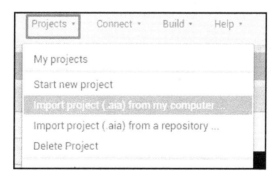

You can download the `.aia` file of the Talking Pi and voice-control bot application file from the `Chapter09` folder of GitHub repository.

Creating a voice-controlled bot app

The voice-controlled bot application is the main focus of this chapter. The following are the three main parts involved in creating a voice-controlled robot:

- **A voice recognition application**: The voice recognition application will recognize our voice and send data as text when a particular word is recognized. For example, if we say the word **forward**, the application would send an **F** to the robot.
- **Bluetooth connection**: This involves establishing a working connection between the Bluetooth of the smartphone and the Bluetooth of the RPi.
- **RPi Robot program**: In this section, we will decode the text information that is transmitted from the smartphone and move the robot accordingly. For example, if the incoming text is **F**, then we will write a program to move the robot forward.

In this section, we will create a voice recognition application. In the later sections, we will look at establishing a Bluetooth connection and programming our robot. You can download the `VoiceControlBot.aia` file from the `Chapter09` folder of GitHub repository.

To create the `VoiceControlBot` app, click on **Projects** and then select **Start new Project**:

Call it `VoiceControlBot` and then press the **OK** button:

Let's now move on to the designing part.

Designing the app

Designing the voice-control bot application is very easy. The final application will look as follows:

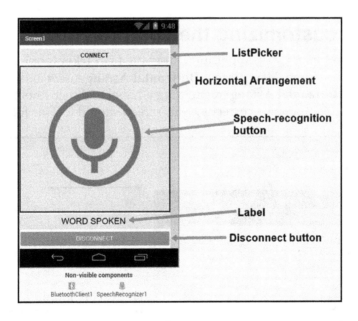

The following components will be used for designing the application:

- **ListPicker**: The ListPicker will display a list of Bluetooth devices that are connected to our smartphone.
- **Speech-recognizer**: The speech-recognizer component will listen to what we are saying.
- **Speech-recognizer button**: When the speech-recognizer button is clicked, the speech-recognizer component will be called, which will listen to what we are saying.
- **Disconnect button**: The disconnect button is used to disconnect the smartphone from the RPi.
- **Label**: The label component will display the text that is spoken by the user.
- **Bluetooth client**: The Bluetooth client component activates our smartphone's Bluetooth connection.
- **Horizontal or vertical arrangement**: We have one horizontal arrangement component to position the speech-recognizer button properly in the center of the screen.

Next let's see how to add and customize components.

Adding and customizing the components

To design the VoiceControlBot application, drag the **ListPicker** (not ListView) component inside the screen. Next, drag a **Horizontal Arrangement** and, inside that, drag a **Button**. Below the **Horizontal Arrangement**, drag a **Label** and then another **Button**. If you have dragged all the components correctly, your screen should look as follows:

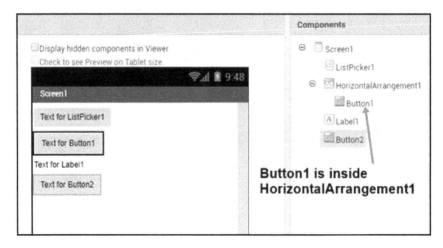

In the following steps, I have customized each component in the app based on my own requirements. You can customize the components as you wish. Take the following steps:

1. First, with the **ListPicker1** selected, change the **background color** to green, set the **Width** to `Fill parent`, and change the **text** to `CONNECT`, as follows:

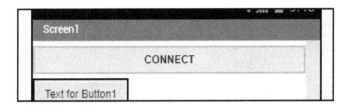

2. Next, select the **HorizontalArrangement1** and change both its **Height** and **Width** to `Fill parent`. Change the **AlignHorizontal** and **AlignVertical** to `Center` so that **Button1** is positioned in the center of the **HorizontalArrangement1**, as follows:

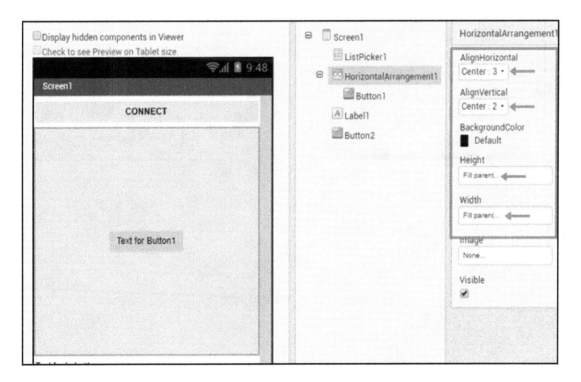

3. After this, select **Button1**, click on the **Rename** button and rename **Button1** to SRButton. **SR** is short for **speech recognition:**

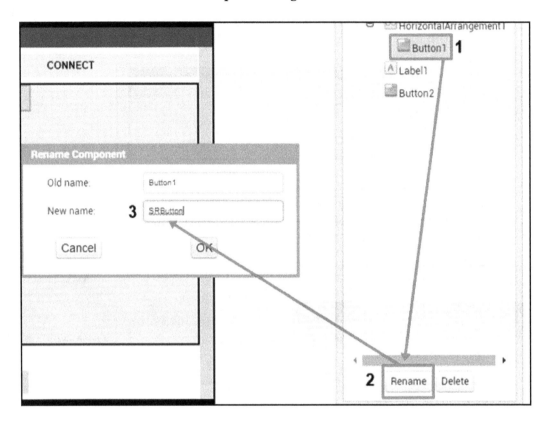

4. Next, we will add a microphone image as the background of the **SRButton**. You can download this image from the `Chapter09` folder of the GitHub repository. To add a background image, change the **Width** and **Height** of the button to `200 pixels`, so that the button is square. Next, remove the default text from the **Text** box, as follows:

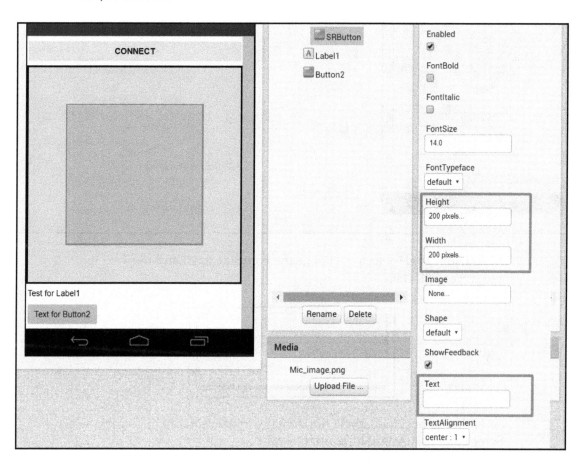

5. After this, click on the **Image** option and then select the mic image to set it as the background image for the **SRButton:**

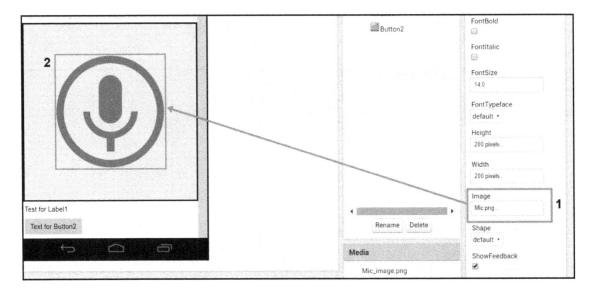

6. Next, with **Label1** selected, change the **FontSize** to 20 and the **Text** to WORD SPOKEN as follows:

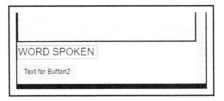

7. After this, to position **Label1** horizontally in the center of the screen, select **Screen1** and set **AlignHorizontal** to Center.

8. Finally, select **Button2** and rename it as `DeleteButton`. Change its **Background color** to RED, its **Width** to `Fill parent`, its **Text** to DELETE, and the **TextColor** to White, as follows:

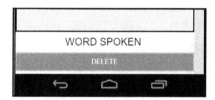

9. After designing the UI for the app, we need to drag the **BluetoothClient** and the **SpeechRecognizer** component to our screen. The **BluetoothClient** is inside the **Connectivity** option and the **SpeechRecognizer** component is inside the **Media** option:

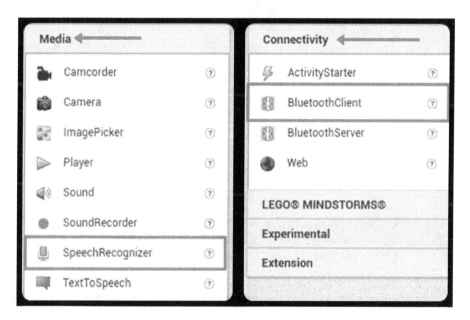

10. Once you have added all the necessary components, your screen should look as follows:

Let's now move on to programming the voice-controlled bot block.

Programming the voice-controlled bot block

After designing the app, it's time to program the voice-controlled bot application:

1. First, we will use the **ListPicker1.BeforePicking** block from **ListPicker1** and display the Bluetooth devices connected to our smartphone in a list as list items. Connect this block to the **ListPicker1.Elements** blocks. Next, from the **BluetoothClient1** components, connect the **BluetoothClient1.AddressAndNames** block to the **ListPicker1.Elements** block, as shown in the following screenshot:

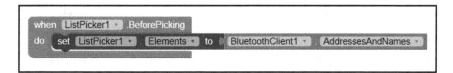

The **ListPicker1.Elements** represents the elements (the list items) in the list, which are the addresses and names of the devices that are paired to our smartphone's Bluetooth. If we select an element from the list, the **ListPicker1.AfterPicking** block comes into play.

2. The **ListPicker1.AfterPicking** block is used to connect to the **AddressesAndNames** Bluetooth, which is selected from the list, as shown in the following screenshot:

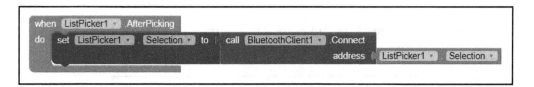

3. Once both the devices are connected using Bluetooth, choose the **SRbutton.Click** block from the **SRbutton**. Then, select the **SpeechRecognizer1.GetText** from the **SpeechRecognizer1** block and connect it to the **SRbutton.Click** block as follows:

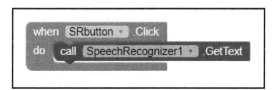

The **SpeechRecognizer1.GetText** block will activate the Google speech recognizer and try to recognize what you are saying. It will convert your spoken words into text.

4. Next, using the **SpeechRecognizer1.AfterGettingText** block, we will display the spoken text inside the label, for example:

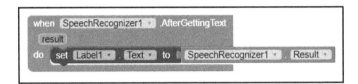

5. After this, using the `if...then` blocks, we will establish whether the spoken words are forward, back, left, right, or stop. If any of these words are detected, then we will send an alphabetical character to our RPi robot using the **BluetoothClient** component. The **if...then** block is inside the **Control** option, as shown in the following screenshot:

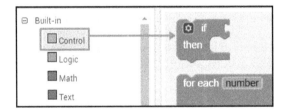

6. Select the **if...then** block and place it below the **Label1.Text** block inside the **SpeechRecognizer1.AfterGettingText** block, as follows:

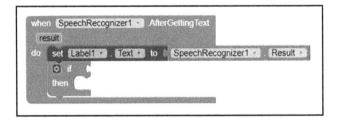

7. Next, to understand the spoken words, we will use the comparison operator, which is inside the **Logic** option, as shown in the following screenshot:

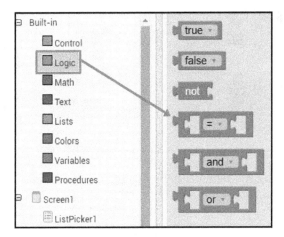

8. Connect the comparison block in the `if` socket of the `if...then` block as follows:

9. In the left socket of the comparison operator, connect the **SpeechRecognizer1.Result** block:

10. In the right socket of the comparison operator, connect an empty text string box. The text string box is inside the **Text** option:

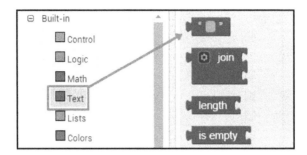

11. After connecting the text string box to the comparison operator, enter the text `forward` inside it as follows:

12. This means that if the **SpeechRecognizer1.Result** is equal to **forward**, then inside the **then** socket, we will add a **BluetoothClient1.SendText** block. After this, we will connect a textbox to the **BluetoothClient1.SendText** block and enter the letter F, as follows:

This means that when the word forward is detected, the character **F** will be sent from the smartphone's Bluetooth to the RPi's Bluetooth.

Right-click on the `if...then` block and duplicate it to create four additional blocks for the words back, left, right, and stop. When the word **back** is detected, you should send the letter **B**; when the word **left** is detected, you should send **L**; when the word **right** is detected, you should send **R**; and, finally, when the word **stop** is detected, you should send **S**. This is shown in the following screenshot:

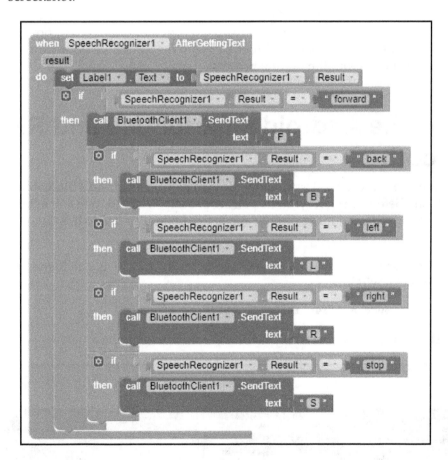

13. After this, connect the **BluetoothClient1.Disconnect** block to the **Disconnectbutton**, so that the Bluetooth connection is disconnected when this button is pressed:

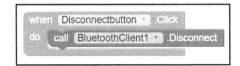

We have now finished designing and programming our `VoiceControlBot` application. You can download and install this application inside your Android smartphone. In the next section, we will pair our smartphone's Bluetooth with the RPi's Bluetooth. Power up your RPi and let's get started with the pairing process.

Pairing the Android smartphone and RPi via Bluetooth

In this section, we will use the Terminal window to pair the Android's Bluetooth with the RPi's Bluetooth. Before we start with the pairing process, we need to install a Bluetooth package inside our RPi and make modifications to certain files. To do this, take the following steps:

1. First, to install the Bluetooth package, enter the following command in the Terminal window:

```
sudo apt-get install libbluetooth-dev
```

The output of the preceding command can be seen in the following screenshot:

```
pi@raspberrypi:~ $ sudo apt-get install libbluetooth-dev
Reading package
Building dependency tree
Reading state information... Done
libbluetooth-dev is already the newest version (5.43-2+rpt2+deb9u2).
0 upgraded, 0 newly installed, 0 to remove and 0 not upgraded.
pi@raspberrypi:~ $
```

2. Next, we will open the `bluetooth.service` file and make some minor modifications. To open the file, enter the following command:

   sudo nano /lib/systemd/system/bluetooth.service

3. Next, type `-C` after `/bluetoothd`. This will turn on the compatible mode of the RPi's Bluetooth as follows:

```
  GNU nano 2.7.4      File: /lib/systemd/system/bluetooth.service

[Unit]
Description=Bluetooth service
Documentation=man:bluetoothd(8)
ConditionPathIsDirectory=/sys/class/bluetooth

[Service]
Type=dbus
BusName=org.bluez
ExecStart=/usr/lib/bluetooth/bluetoothd -C
NotifyAccess=main
#WatchdogSec=10
#Restart=on-failure
CapabilityBoundingSet=CAP_NET_ADMIN CAP_NET_BIND_SERVICE
LimitNPROC=1
```

4. After this, press *Ctrl + O* and then hit *Enter* to **save** the file. Next, press *Ctrl + X* to exit the file. Reboot your RPi with the following command:

   sudo reboot

5. Once the RPi has restarted, check its status by entering the following command:

   sudo service bluetooth status

You should now see -C next to `bluetoothhd`. If, after entering the preceding command, you see `lines 1-19/19(END)` and you can't enter any new commands in the Terminal window, close the Terminal window and reopen it again:

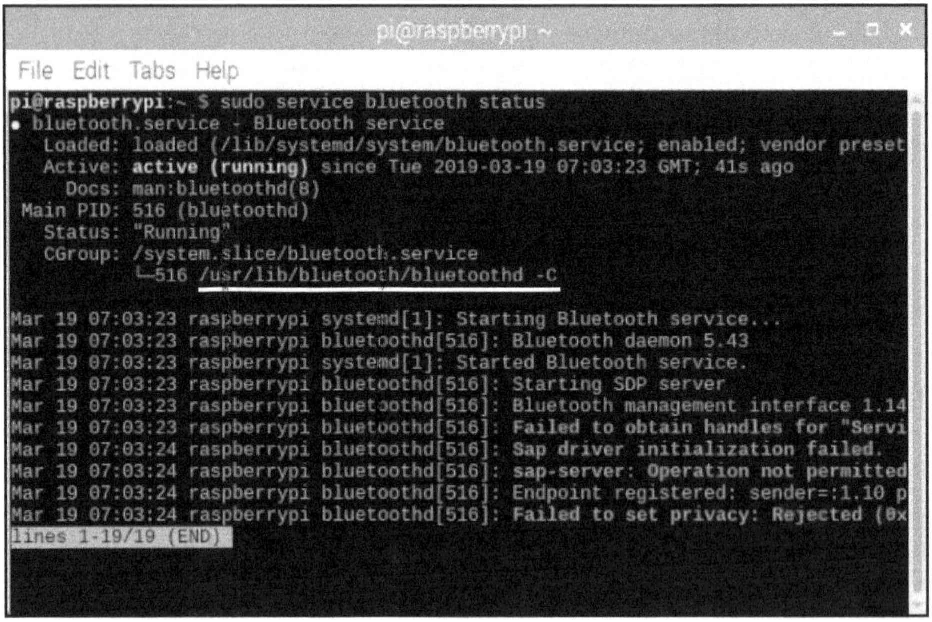

6. Next, to pair the RPi's Bluetooth with the smartphone's Bluetooth, we first need to make the RPi's Bluetooth discoverable. To pair the devices, enter the following command:

```
sudo bluetoothctl
```

7. You should now see the **Media Access Control** (**MAC**) address of your Bluetooth along with its name. The MAC address is a 12-digit address and is unique for each Bluetooth device. Your RPi's Bluetooth will have a default name of `raspberrypi`, as you can see in the following screenshot:

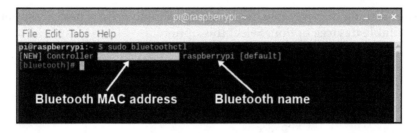

Bluetooth MAC address   **Bluetooth name**

8. Before entering the next set of code, **turn your Android device's Bluetooth on** and click on the **Pair new device** option as follows:

9. After this, enter the following five commands one by one to put the RPi's Bluetooth in discoverable mode:

```
power on       //turns on the Bluetooth

pairable on    //Bluetooth is ready to pair with other Bluetooth

discoverable on   //Bluetooth is now in discoverable mode

agent on       //Bluetooth agent is the one which
               //manages Bluetooth pairing
               //code. It can respond to incoming pairing
               //code and it can also
               //send out pairing code
default-agent
```

10. After turning discoverable mode on, you should see the name **raspberrypi** in the **Available devices** option. Select this option:

11. After selecting the **raspberrypi** option, you will see a pop-up box asking you whether you want to pair with the Raspberry Pi's Bluetooth. Click on the **PAIR** button:

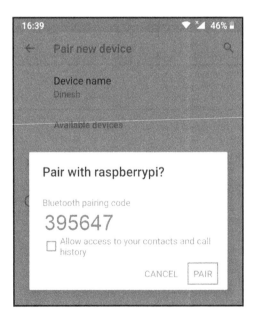

12. Next, inside the Terminal window, you will see a message asking whether you want to pair with your smartphone's Bluetooth. Type `yes` (in lower-case letters) and press *Enter*:

13. Next, you will see a small pop-up box in the Terminal window asking if you want to accept the pairing request. Click **OK**:

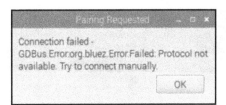

14. You may then see another pop-up box stating that the connection failed. Ignore this error and press **OK**:

15. After this, type `exit`. You can now check whether the device is paired to your RPi by typing `sudo bluetoothctl`, as follows:

```
[bluetooth]# exit
Agent unregistered
[DEL] Controller B8:27:EB:BE:D3:C0 raspberrypi [default]
pi@raspberrypi:~ $ sudo bluetoothctl
[NEW] Controller B8:            raspberrypi [default]
[NEW] Device 6C:            Dinesh
[bluetooth]#
```

So we have finished pairing the RPi's Bluetooth to the Android smartphone's Bluetooth. Next we will enable the serial port of the RPi Bluetooth.

Enabling the Bluetooth serial port

After pairing the devices, we need to create a script for enabling the Bluetooth serial port. We will name this script `bt_serial` (**bt** being short for **Bluetooth**). To create this script, follow these instructions:

1. Type the following command:

   sudo nano bt_serial

2. Inside this script, enter the following lines:

   ```
   hciconfig hci0 piscan
   sdptool add SP
   ```

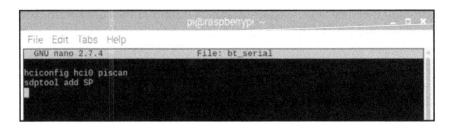

3. Next, save (*Ctrl* + *O*) and exit (*Ctrl* + *X*) this script.

4. We then need to execute and run this script. Type the following commands: `sudo chmod +x bt_serial` (this is the execution command) and `sudo ./bt_serial` (this is the run command):

After running the script, you will see the message `Serial Port service registered`.

Developing the Bluetooth program for RPi

After designing the Android application, pairing the devices, and enabling the serial port, it is now time to program the RPi so that it can receive text data from the Android smartphone. To receive incoming text data from the smartphone, we are going to use sockets from socket programming.

Socket programming

A socket is an endpoint of a two-way communication system in a network. We create sockets so that we can send bits of information through them. To establish a Bluetooth communication between the devices, we need to create a socket. One socket will be on the server side and the other will be on the client side. In our case, the Android smartphone is the client and the RPi is the server.

The client socket tries to establish a connection with the server socket, while the server socket tries to listen to the incoming connection request from the client socket. In Bluetooth programming, we can choose between two socket options, RFCOMM and L2CAP, as represented in the following diagram:

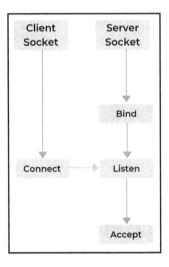

In a socket program, the following connection steps have to happen one after another from the client and server side. Each step represents a function, which is declared either in the client, server, or both scripts. Take the following steps:

1. **Socket creation (client/server)**: Creates sockets in the client and server programs as follows:

   ```
   socket(int domain, int type, int protocol)
   ```

   In this function, the first parameter refers to the communication domain. In our case, the communication domain is Bluetooth (`AF_BLUETOOTH`). The second parameter refers to the communication type (`SOCK_STREAM`). The third parameter refers to the communication protocol. In Bluetooth programming, we can choose the **Radio Frequency Communication (RFCOMM)** protocol or the **Logical Link Control and Adaption Protocol (L2CAP)**.

2. **Connect (client)**: This function tries to establish a connection with the server socket:

   ```
   connect(int sock, const struct sockaddr *server_address,
   socklen_t info)
   ```

In this function, the first parameter refers to the socket, the second parameter refers to the address of the server, and the third parameter is used to find the size of device address. In our case, the device address is the RPi address.

3. **Bind (server)**: The `bind` function binds the address and port number of the server device. The MAC address of the RPi will be stored inside the bind function as follows:

```
bind(int sock, const struct sockaddr *address, socklen_t info);
```

In this function, the first parameter refers to the socket, the second parameter refers to the address of the server device (Raspberry Pi), and the third parameter is used to find the size of the device address.

4. **Listen (server)**: With the `listen` function, the server socket waits for the client to approach it in order to make a connection:

```
listen(int sock, int backlog);
```

The first parameter refers to the socket. The backlog is generally set to 1.

5. **Accept (server)**: The `accept` function waits for an incoming connection request and creates a new socket:

```
int new_socket = accept(int sock, struct sock_address
*clientaddress, socklen_t info);
```

In this function, the second parameter refers to the address of the client, which is the Android smartphone.

6. **Send (client/server)**: The `send` function is used to send data from the client to the server and the other way around.
7. **Read (client/server)**: The `read` function is used to read data transferred from the client to the server and the other way around.
8. **Close (client/server)**: The `close` function shuts down the socket and frees the memory allocated to the socket.

Now, since we have already created the `VoiceControlBot` Android application using App Inventor, there is no need to write a client program. All that's left is to write the server program for our RPi robot.

VoiceBot server program

In this `VoiceBot` server program, we will first use a socket program to establish a socket connection between the devices. Next, we will receive the incoming data, which is sent from the Android smartphone. Finally, we will move the robot based on the data that is sent. The `VoiceBot.c` program can be downloaded from the `Chapter09` folder of the GitHub repository. Follow these steps:

1. First, all the necessary header files are declared as follows:

```
#include <stdio.h>
#include <unistd.h>
#include <sys/socket.h>                //Socket header file
#include <bluetooth/bluetooth.h>       //Bluetooth header file
#include <bluetooth/rfcomm.h>          //Radio frequency
communication header file
#include <wiringPi.h>
```

2. Next, inside the `main()` function, the wiringPi pin numbers 0, 2, 3, and 4 are declared as output pins as follows:

```
pinMode(0,OUTPUT);
pinMode(2,OUTPUT);
pinMode(3,OUTPUT);
pinMode(4,OUTPUT);
```

To establish RFCOMM with the other device, `sockaddr_rc` is declared along with the `server_address` and `client_address`. The `data` variable will store and display the incoming data. The `s` and `clientsocket` are for storing the values of the server and client socket respectively. The `bytes` variable will read the incoming byte information. The `socklen_t opt` contains the size of `client_address`. This is shown in the following lines of code:

```
struct sockaddr_rc server_address = { 0 }, client_address = { 0
};
char data[1024] = { 0 };
int s, clientsocket, bytes;
socklen_t opt = sizeof(client_address);
```

3. Next, the `s` is created with the communication domain set to `AF_BLUETOOTH`, the communication type set to `SOCK_STREAM`, and the communication protocol set to `BTPROTO_RFCOMM`:

```
s = socket(AF_BLUETOOTH, SOCK_STREAM, BTPROTO_RFCOMM);
```

4. The RPi's Bluetooth MAC address is then bound to the server socket using the bind function and the s (server socket) enters listening mode as follows:

```
bind(s, (struct sockaddr *)&server_address,
sizeof(server_address));
listen(s, 1);
```

5. Once the connection is accepted, a new client socket is created using the accept function as follows:

```
clientsocket = accept(s, (struct sockaddr *)&client_address,
&opt);
```

6. The incoming byte data is then converted to a string using the ba2str function. After that, the MAC address of the connected Bluetooth is displayed as follows:

```
ba2str( &client_address.rc_bdaddr, data );
fprintf(stderr, "Connected to %s\n", data);
```

7. After this, inside the for loop, the incoming data is read using the read function. If the value inside the byte variable is greater than 0, we print the data as follows:

```
for(;;){
bytes = read(clientsocket, data, sizeof(data));
if( bytes > 0 ) {
printf("Alphabet: %s\n", data);
```

Now, we will use five if conditions to check whether the incoming data contains the letters F, B, L, R, or S, as follows:

```
if(*data=='F')
{
----Forward Code----
}
else if(*data=='B')
{
----Backward Code----
}
else if(*data=='L')
{
----Axial Left Turn Code----
}
else if(*data=='R')
{
----Axial Right Turn Code----
}
else if(*data=='S')
```

```
{
----Stop Code----
}
```

The preceding code can be explained as follows:

- `*data == 'F'`: The robot will move forward
- `*data == 'S'`: The robot will move backward
- `*data == 'L'`: The robot will take an axial left turn
- `*data == 'R'`: The robot will take an axial right turn
- `*data == 'S'`: The robot will stop

8. Finally, to disconnect the devices, the `close` function is used to close the `clientsocket` and the `s` as follows:

```
close(clientsocket);
close(s);
```

9. Next, since this code contains a socket and a Bluetooth header file, to compile this code, you will need to add the `-lbluetooth` command inside the **Build** command. As this is a C program and not a C++ program, you will also have to add the **-lwiringPi** command to compile the wiringPi code as follows:

Next let's test the code and check the final output.

Testing the code

Now, before compiling and building the program, make sure that you have paired both the Android smartphone's Bluetooth and the RPi's Bluetooth. If they are not paired, then the RPi Bluetooth name (`raspberrypi`) will not appear in the Bluetooth list when you run the `VoiceControlBot` application. The steps for pairing the devices are listed in the *Pairing the Android smartphone and RPi via Bluetooth* section.

Once you have done this, you will need to execute and run the `bt_serial` script, which we created previously, inside the Terminal window. The commands for executing and running this script are as follows:

```
sudo chmod +x bt_serial          //Execution code
sudo ./bt_serial                 //Run Code
```

You do not need to execute this script each time you run the program but you will need to execute and run this script when you start a new RPi session and want to test the `VoiceBot.c` program. Next, compile and build the `VoiceBot.c` program. After this, open the `VoiceControlBot` Android app and press the **CONNECT** list picker. You will see the name of the **raspberrypi** along with its MAC address in the Bluetooth list. Select the **raspberrypi** option to connect the devices as follows:

Once they are connected, you will get a notification inside the Terminal window stating `Connected to:` and the Android Bluetooth MAC address, as shown in the following screenshot:

If you get the following **Error 507: Unable to Connect. Is the device turned on?** error, don't worry. Click on the connect button and select the **raspberrypi** Bluetooth again:

Once the devices are connected, you can click on the speech recognizer button and start speaking. If you say the word *forward*, this should be displayed on the screen, as in the following screenshot, and the letter **F** will be sent to the RPi:

Similarly, when you say the words *back*, *left*, *right*, and *stop*, the letters **B**, **L**, **R**, and **S**, as shown in the following screenshot, will be sent to the RPi's Bluetooth and the robot will move in the appropriate direction:

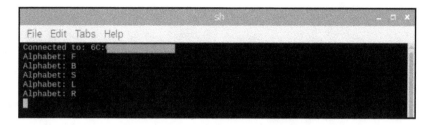

If you say any other word, the speech recognizer should recognize the word and display it on the screen, but it will not send any text data to the RPi.

Summary

We started this chapter by creating our first Android application called Talking Pi, in which text written inside the textbox was displayed in a label and also read out by the smartphone. We then developed a voice-controlled bot Android app, which recognized our voice and sent text to the RPi via Bluetooth. After this, using the Terminal window, we paired the Android smartphone's Bluetooth with the RPi's Bluetooth. Finally, we looked at socket programming and wrote the `VoiceBot` program to establish a connection with the Android smartphone's Bluetooth to control the robot.

Questions

1. Which communication protocol did we use to send data over a Bluetooth connection?

2. What kind of address do Bluetooth devices have?

3. What was the use of ListPicker inside the `VoiceControlBot` application?

4. On the client side, which function is used to connect the client socket to the server socket?

5. What is the default Bluetooth name of your RPi?

Assessments

Chapter 1: Introduction to the Raspberry Pi

1. A Broadcom BCM2837 quad-core 1.4 GHz processor
2. 40
3. VNC Viewer
4. Username: `pi`, Password: `raspberry`
5. `sudo raspi-config`

Chapter 2: Implementing Blink with wiringPi

1. Eight (pin numbers **6**, **9**, **14**, **20**, **25**, **30**, **34**, and **39**)
2. HIGH
3. `digitalRead(pinnumber);`
4. `for (int i=0; i<6;i++)`
5. 1V

Chapter 3: Programming the Robot

1. L298N motor driver IC
2. H-bridge
3.
```
digitalWrite(0,HIGH);
digitalWrite(2,LOW);
digitalWrite(3,HIGH);
digitalWrite(4,LOW);
```

4. Anticlockwise direction
5.
```
digitalWrite(0,HIGH);
digitalWrite(2,HIGH);
digitalWrite(3,HIGH);
digitalWrite(4,LOW);
```

6. ```
digitalWrite(0,HIGH);
digitalWrite(2,LOW);
digitalWrite(3,LOW);
digitalWrite(4,HIGH);
```

# Chapter 4: Building an Obstacle-Avoiding Robot

1. An ultrasonic pulse traveling at 340 m/s
2. Liquid crystal display
3. 180 cm
4. Column 4 and row 1
5. They are used to adjust the LCD's backlight

# Chapter 5: Controlling a Robot Using a Laptop

1. `initscr()` and `endwin()`
2. The `initscr()` function initializes the screen. It sets up the memory and clears the command window screen
3. `gcc -o Programname-lncurses Programname.cpp`
4. GCC
5. `pressed()` to move the robot while the button is pressed, and `released()` to stop it

# Chapter 6: Building an Object-Following Robot with OpenCV

1. Open Source Computer Vision
2. 3,280 x 2,464 pixels
3. `raspistill`
4. `raspivid`
5. 8GB - 50% and  32 GB - 15%
```

Chapter 7: Accessing the RPi Camera with OpenCV

1. Thresholding
2. `flip(original_image, new_image, 0)`
3. The bottom-right block
4. `resize(original_image , resized_image , cvSize(640,480));`
5. On the top-left part of the screen

Chapter 8: Face Detection and Tracking Using the Haar Classifier

1. `haarcascade_frontalface_alt2.xml`.
2. Horizontal Line feature.
3. `haarcascade_lefteye_2splits.xml`.
4. Region of Interest.
5. The `equalizeHist` function improves the brightness and contrast of the image. This is important because in low lighting, the camera may not be able to distinguish the face from the image.

Chapter 9: Building a Voice-Controlled Robot

1. **Radio Frequency Communication (RFCOMM)**
2. **Media Access Control (MAC)** address
3. The ListPicker displayed a list of all Bluetooth devices that were already paired with your smartphone's Bluetooth
4. Connect
5. `raspberrypi`

Other Books You May Enjoy

If you enjoyed this book, you may be interested in these other books by Packt:

Learn Robotics Programming
Danny Staple

ISBN: 978-1-78934-074-7

- Configure a Raspberry Pi for use in a robot
- Interface motors and sensors with a Raspberry Pi
- Implement code to make interesting and intelligent robot behaviors
- Understand the first steps in AI behavior like speech recognition visual processing
- Control AI robots using Wi-Fi
- Plan the budget for requirements of robots while choosing parts

Mastering ROS for Robotics Programming
Lentin Joseph

ISBN: 978-1-78355-179-8

- Create a robot model of a Seven-DOF robotic arm and a differential wheeled mobile robot
- Work with motion planning of a Seven-DOF arm using MoveIt!
- Implement autonomous navigation in differential drive robots using SLAM and AMCL packages in ROS
- Dig deep into the ROS Pluginlib, ROS nodelets, and Gazebo plugins
- Interface I/O boards such as Arduino, Robot sensors, and High end actuators with ROS
- Simulation and motion planning of ABB and Universal arm using ROS Industrial
- Explore the ROS framework using its latest version

Leave a review - let other readers know what you think

Please share your thoughts on this book with others by leaving a review on the site that you bought it from. If you purchased the book from Amazon, please leave us an honest review on this book's Amazon page. This is vital so that other potential readers can see and use your unbiased opinion to make purchasing decisions, we can understand what our customers think about our products, and our authors can see your feedback on the title that they have worked with Packt to create. It will only take a few minutes of your time, but is valuable to other potential customers, our authors, and Packt. Thank you!

Index